Introductory Physics for Engineers

Introductory Physics for Engineers

By

Waleed S. Mohammed
and Pimrapat Thanusutiyabhorn

Cambridge
Scholars
Publishing

Introductory Physics for Engineers

By Waleed S. Mohammed and Pimrapat Thanusutiyabhorn

This book first published 2024

Cambridge Scholars Publishing

Lady Stephenson Library, Newcastle upon Tyne, NE6 2PA, UK

British Library Cataloguing in Publication Data
A catalogue record for this book is available from the British Library

ISBN: 978-1-0364-1144-2
ISBN (Ebook): 978-1-0364-1145-9

To the children of Gaza

"The duty of the man who investigates the writings of scientists, if learning the truth is his goal, is to make himself an enemy of all that he reads, and, applying his mind to the core and margins of its content, attack it from every side."

Form The book of optic
Al-Hassan Ibn-El-Haytham
965-1040 AD

TABLE OF CONTENTS

PREFACE

The provided textbook focuses on basic concepts that covers topics of Newton mechanics, oscillation, electromagnetics, fluidic and thermodynamics. It elaborates the three laws of motions as well as the laws of conservation of energy and momentum. It then links the laws of motion to oscillating movement. The concept of oscillation is used to explain electrical oscillators and electromagnetic radiation. The book then specified two chapters to give brief introduction to fluid systems and thermodynamics. That includes the definition of fluid and the basics of fluid mechanics that include static systems as well as fluid dynamics. Fundamental concepts such as heat, entropy, heat transfer and thermal engines are covered.

While writing the book, the authors kept in mind that it is targeting first year undergraduate engineering students. Hence, we find it necessary to start the book with a brief mathematical review to cover some needed background such as vectors, vector operations, trigonometry, differentiation and integration. That is followed by introducing scalar and vector quantities such as displacement, velocity and acceleration. The chapter is made to provide a brief revision material for students who might not have covered all or some these topics in mathematics during their pre-university education.

Additionally, during the course of the book, the authors were careful to ensure that level and use of mathematics gradually increase. For each concept introduced, full mathematical derivations are provided for the governing equations. This might seem to mathematically challenge some student groups. However, all of the needed mathematical operations are briefly covered in the first chapter.

—The authors

CHAPTER ONE

INTRODUCTION TO MOTION

1A. Vectors brief review

When dealing with topics such as motion and Newton's mechanics, it is essential to build a good understanding of vectors and operations. This is because motion within the limits of Newton's mechanics can be oriented in a three-dimensional space. Each dimension requires a value to describe a location or the change of that location along this dimension. Hence, being in a three-dimensional space requires three components to describe the motion through such space. Once any quantity requires more than one component to be represented, we could fundamentally call it a vector quantity.

Distance and displacement vector

Let us imagine the following scenario:

A student wanted to plan a trip from Chiang Mai to Bangkok. Being new to Thailand, he asked how far is it to travel from Chiang Mai to Bangkok? A local student answered quickly, "it is 700 km (about 435 mi)". The first student thought for a moment then replied, Which direction? " As a teacher, you decided to interrupt without a doubt, "South?" The first student, being a sharp kid, replied, "Is it pointing directly to the south?" You were about to give him a fast answer looking at the map in figure 1.1 "South-East", but then you asked an important question:

"How much south and how much east?"

To reach a solution, you found yourself prone to set a coordinate system originated at Chiang Mai where x matches East and y matches North. Now, you can draw a 700 km long arrow that points directly from Chiang Mai to Bangkok. This arrow then forms a vector: Displacement vector of 700 km length and a direction as shown in figure 1.2.

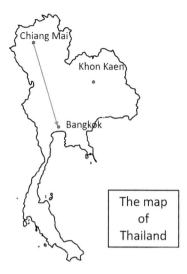

Figure 1.1. The map of Thailand at which the students are looking at. The map shows an arrow pointing from Chiang Mai to Bangkok. The length of the arrow is 700 km.

Vector representation

One can also say, to reach from Chiang Mai (point A) to Bangkok (Point B) you would need to move 669 km (about 415.7 mi) South then 204 km (about 126.76 mi) East. If we placed a two-dimensional coordinate system where North is along the y-axis and x-axis is along East, as in figure 1.3, then the South should have a negative sign. Or we simply say, move − 669 km (about 415.7 mi) on the y-axis, then move 204 km (about 126.76 mi) on the x-axis. In a vector form one writes the displacement vector as:

$$\overline{L} = (240 km, -669 km) \tag{1.1}$$

For this vector can be defined by

- Amplitude: $|\overline{L}| = \sqrt{204^2 + 669^2} = 700$ km.
- Direction: $\theta = tan^{-1}(-669/204) = $ -73°.

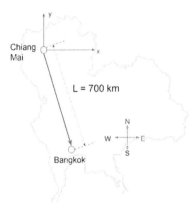

Figure 1.2. A displacement vector between two points: Chiang Mai and Bangkok. The vector has: A length that equals the distance between the two points and a direction pointing from the starting point to the ending point.

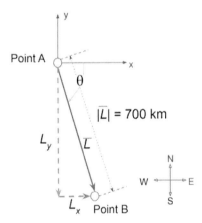

Figure 1.3. The displacement vector representation by a length L and angle θ that represents direction.

A general displacement vector can be written as

$$\overline{L} = (L_x, L_y) \tag{1.2}$$

The vector amplitude is defined as

$$|\overline{L}| = \sqrt{L_x^2 + L_y^2} \tag{1.3}$$

The vector direction is defined as

$$\theta = tan^{-1}(L_y/L_x) \tag{1.4}$$

For the general vector in figure 1.4, with length U and angle θ, we can write it as a summation of two vectors

$$\overline{u} = u_x\hat{x} + u_y\hat{y}. \tag{1.5}$$

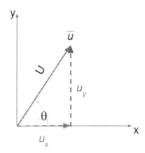

Figure 1.4. General vector of amplitude U and direction θ.

The vectors $\hat{x}$ and $\hat{y}$ are unit vectors in the x and y directions. They each have an amplitude of one. The vector can be then written in the form

$$\overline{u} = (u_x, u_y) \tag{1.6}$$

where u_x and u_y are the x and y components of the vector $\overline{u}$. We can obtain their values from the vector length U using Pythagoras theorem as follows.

$$u_x = U \cdot cos\theta \tag{1.7a}$$

$$u_y = U \cdot sin\theta \tag{1.7b}$$

The vector's direction is defined by the angle it makes with respect to the x-axis. From the geometry in figure 1.4, we can immediately realize that the ratio between u_y and u_x is the tangent of the angle.

$$\frac{u_y}{u_x} = \frac{sin\theta}{cos\theta} = tan\theta \tag{1.8}$$

Vector and scalar

In the first example, the vector given was the displacement vector. The amplitude of the vector is a scalar value that indicates the distance measured as a straight line between the starting and ending points. The vector's direction is represented by an angle to the x-axis formed between the starting point and the end point. The angle itself is a scalar value as it does not have a direction.

Vector operations

Back to the map. Imagine that there is a salesperson who travels from Chiang Mai to Bangkok then to Khon Kaen. What would be his net displacement? From the definition in the previous sections, we know that the displacement is a vector that is formed between the starting and ending points of the trip. In this case, the starting point is Chiang Mai, and the ending point is Khon Kaen

How do we represent this in vectors?

Figure 1.5. The salesman trip from Chiang Mai to Bangkok then to Khon Kaen and the net displacement vector in dashed arrow.

The total displacement can then be represented as a sum of the two displacements that the salesperson did. The first is from point A→B and the second is from B→C. Mathematically we write this as the following vector summaries

$$\overline{AC} = \overline{AB} + \overline{BC} = \left(AB_x, AB_y\right) + \left(BC_x, BC_y\right) = \left(AB_x + BC_x, AB_y + BC_y\right)$$
$$(1.9)$$

In our example AB_y is pointing south and hence has a negative sign, while BC_y is pointing north which makes it positive. The summation $AB_y + BC_y$ is the subtraction of the two amplitudes and the result is a shorter distance as illustrated in figure 1.6. The x components however, AB_x and BC_x are both pointing to the east. Hence, the summation gives us the addition of the two amplitudes. In general, adding two vector quantities is simply the summation of each corresponding component individually, e.g., adding the x component of each vector to form the x component of the final vector. One can do similar proof for vector subtraction using a similar example. Though we can write the two operations as follows:

$$\overline{u} + \overline{v} = (u_x + v_x, u_y + v_y) \qquad (1.10)$$

$$\overline{u} - \overline{v} = (u_x - v_x, u_y - v_y) \qquad (1.11)$$

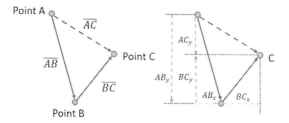

Figure 1.6. Displacement vector between point A and C as sum of two displacements $\overline{AC} = \overline{AB} + \overline{BC}$.

When multiplying a vector u by a scalar b, we simply multiply each component by the scalar.

$$b \cdot \overline{u} = (b \cdot u_x, b \cdot u_y) \qquad (1.12)$$

$$b \cdot (\overline{u} + \overline{v}) = b \cdot \overline{u} + b \cdot \overline{v} \qquad (1.13)$$

Dot product of vectors

So, what about vector-vector multiplication? There are two types of vector multiplication: One that returns a scalar value, dot product, and another returns a vector, cross product. We will be presenting here only the dot product and cross product will be introduced in chapter 9. Dot product returns a scalar value, and it is defined mathematically as

$$\bar{u} \cdot \bar{v} = u_x v_x + u_y v_y \qquad (1.14)$$

If we assume the two vectors have lengths of U and V respectively, then we can use Pythagoras theorem to write the vectors as $\bar{u} = U \cdot cos\theta_u \hat{x} + U \cdot sin\theta_u \hat{y}$ and $\bar{v} = V \cdot cos\theta_v \hat{x} + V \cdot sin\theta_v \hat{y}$. Using this representation, we can write equation 1.14 as

$$\bar{u} \cdot \bar{v} = U \cdot V \cdot (\cos\theta_u \cos\theta_v + \sin\theta_u \sin\theta_v) \qquad (1.15)$$

To simplify the expression inside the parentheses one needs to revisit some basics of trigonometry in the next section.

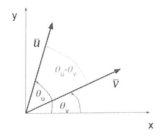

Figure 1.7. Dot product of two vectors

1B. Mathematical review

Brief review of trigonometry

For now, we need to remember the following two relations by heart. This will make derivations much easier when we try to simplify the dot product expression in equation 1.15b.

$$cos\theta = \frac{1}{2}\left(e^{i\theta} + e^{-i\theta}\right) \tag{1.16a}$$

$$sin\theta = \frac{1}{2i}\left(e^{i\theta} - e^{-i\theta}\right) \tag{1.16b}$$

We can use equations 1.16 to expand the sines and cosines in equation 1.15b.

$$sin\theta_u sin\theta_v = \frac{1}{2i}\left(e^{i\theta_u} - e^{-i\theta_u}\right) \cdot \frac{1}{2i}\left(e^{i\theta_v} - e^{-i\theta_v}\right) \tag{1.17a}$$

$$= \frac{-1}{4}\left(e^{i(\theta_u+\theta_v)} + e^{-i(\theta_u+\theta_v)} - e^{i(\theta_u-\theta_v)} + e^{-i(\theta_u-\theta_v)}\right) \tag{1.17b}$$

We can arrange the equations as

$$sin\theta_u \cdot sin\theta_v = \frac{1}{4}\left(e^{i(\theta_u+\theta_v)} + e^{-i(\theta_u+\theta_v)}\right) - \frac{1}{4}\left(e^{i(\theta_u-\theta_v)} + e^{-i(\theta_u-\theta_v)}\right) \tag{1.18}$$

Comparing the two parentheses in equation 1.18 to equations 1.16, we can write the following relation

$$sin\theta_u \cdot sin\theta_v = \frac{1}{2}(cos(\theta_u + \theta_v) - cos(\theta_u - \theta_v)) \tag{1.19}$$

We can follow a similar approach to write the following expression

$$cos\theta_u \cdot cos\theta_v = \frac{1}{2}(cos(\theta_u + \theta_v) + cos(\theta_u - \theta_v)) \tag{1.20}$$

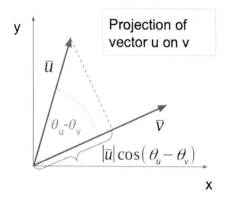

Figure 1.8: projection of $\overline{u}$ on $\overline{v}$

Adding equations 1.19 and 1.20 we obtain the following relation

$$sin\theta_u \cdot sin\theta_v + cos\theta_u \cdot cos\theta_v = cos(\theta_u - \theta_v) \qquad (1.21)$$

Dot product as vector projection

If we substitute equation 1.21 into 1.15b we obtain a more known representation of the vector dot product.

$$\overline{u} \cdot \overline{v} = UV \cdot cos(\theta_u - \theta_v) \qquad (1.22)$$

Or in a general form

$$\overline{u} \cdot \overline{v} = |\overline{u}||\overline{v}|cos(\theta_u - \theta_v) \qquad (1.23)$$

Notice here we write the length of the vector, or amplitude, as $|\overline{u}|$. Equation 1.23 represents the projection of one vector on the other as illustrated in figure 1.8. We can write the projection of the vectors as follows:

- Projection of u on v : $\overline{u} \cdot \overline{v}/|\overline{v}| = |\overline{u}|cos(\theta_u - \theta_v)$
- Projection of v on u : $\overline{u} \cdot \overline{v}/|\overline{u}| = |\overline{v}|cos(\theta_u - \theta_v)$

Review of complex numbers

Generally, a complex number z is defined as $z = a + ib$. Here, a and b are real values. The symbol i is for imaginary and it has a numerical value of $i = \sqrt{-1}$. Complex numbers can be represented in a way similar to vector representation. However, instead of x and y coordinates, we set the real and imaginary coordinates as shown in figure 1.9. The amplitude of the complex number is:

$$|z| = \sqrt{a^2 + b^2} \tag{1.24}$$

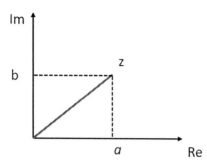

Figure 1.9. Complex number vector representation.

As in figure 1.9 and in a similar way to vector representation, we can write the real and imaginary parts of the vector in terms of cosine and sine functions as in equations 1.17.

$$a = |z|cos\phi \tag{1.25a}$$

$$b = |z|sin\phi \tag{1.25b}$$

Here, the angle $\phi = tan^{-1}(b/a)$. The complex number z can be written then as

$$z = |z|(cos\phi + i\ sin\phi) \tag{1.26}$$

The term in the parentheses in equation 1.26 can be written in an exponential form

$$cos\phi + i\ sin\phi = e^{i\phi} \tag{1.27}$$

The complex number z can be written as

$$z = |z|e^{i\phi} \tag{1.27b}$$

The constant ϕ is commonly called the phase.

Derivative, change and rate

Derivative means change and change can be defined as the difference between the conditions at two different states. The two different states can be two different times, locations in space, or two different temperatures for example. Let us consider the following scenario. You decided to put a thermometer inside an oven and started to record the measured temperature every 10 seconds. From the records, you managed to produce the graph in figure 1.10. Keep in mind that this graph is hypothetical and does not represent actual oven behavior.

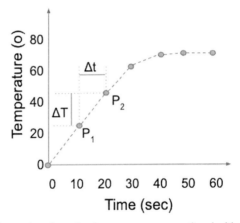

Figure 1.10: The produced graph of temperature versus time inside the oven.

Now, let us pick two points on the graph: P_1 and P_2. As shown by the values in table 1.3, between points P_1 and P_2, the temperature has increased by

$$\Delta T = 45^o - 22^o = 23^o$$

and the time has increased by

$$\Delta t = 20s - 10s = 10s.$$

Hence, we can say that the rate of change between the two points is 23 degrees in 10 seconds.

Table 1.3: Time and temperature values of two points in figure 1.10

Point	Time (s)	Temperature (°)
P_1	10	22
P_2	20	45

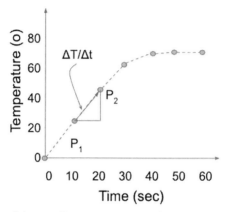

Figure 1.11: Rate of change of temperature versus time.

The rate of change can be defined as

$$Rate = \frac{\Delta T}{\Delta t} = \frac{\text{Difference in temperature}}{\text{Difference in time}} \qquad (1.28)$$

According to equation 1.28, The unit of Rate is degrees per second. For example, for the values here, the rate of change in the oven's temperature between the two points is

$$Rate = \frac{23^o}{10s} = 2.3 \text{ Degrees/sec} \qquad (1.29)$$

Notice, that if you measure while cooling the oven, then the difference in temperature would be negative. That will give you a negative Rate. Now you have decided to connect the thermometer with a very fast computer that reads temperature values with a very small amount of time difference. In other words, the time intervals, Δt, became very small or infinitesimally small. In this condition we can replace the phrase "difference" by differentiation or

When $\Delta T \rightarrow dT$ then $\Delta t \rightarrow dt$ $\hspace{3cm}$ (1.30)

The Rate is now written as

$$Rate = \frac{dT}{dt} = \frac{\text{Differentiation in temperature}}{\text{Differentiation in time}} \hspace{1.5cm} (1.31)$$

In other words, we could say that the rate is the derivative of temperature with respect to time when we use infinitesimally small-time steps.

Functions

With very small-time intervals, one can say that the temperature can "almost" be estimated at any time value. We can think of the graph as the temperature represented in a continuous form rather than a set of discrete points. The graph in figure 1.12 shows the temperature as a continuous function of time. Mathematically, when we want to describe temperature as a function of time, we write T(t). That reads as temperature at time t. How can we now think of the rate of change of a continuous function?

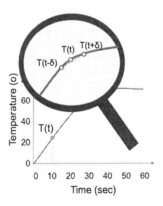

Figure 1.12: Rate of change of temperature versus time.

We know that for very small intervals the rate of change was defined as the ratio of the differentiation of temperature over the differential of time. This was, however, calculated between two discrete points. In a continuous form we have a function T(t) that gives us the temperature for each time value. To get a better understanding of that, we need to zoom at a selected region in the continuous graph as shown in figure 1.12. We can think that the rate of change around a time t is estimated as the difference in temperature just before t and just after t divided by the small step of time change.

$$Rate = \frac{dT(t)}{dt} \qquad (1.32a)$$

$$dT(t) = T(t + \delta) - T(t - \delta) \qquad (1.32b)$$

$$dt = t + \delta - (t - \delta) = 2\delta \qquad (1.32c)$$

The time step δ is infinitesimally small or it's value approaches zero, $\delta \to 0$. The rate of change in equation 1.32a, which is the derivative of temperature with respect to time can be written in a approximate form following equations 1.32b and 1.32c as

$$Rate = \frac{dT(t)}{dt} = \lim_{\delta \to 0} \frac{T(t+\delta) - T(t-\delta)}{2\delta} \qquad (1.33)$$

Example 1.1: Derivative of a linear function

A linear function is typically written as

$$T(t) = a + b \cdot t \qquad (1.34)$$

Here a and b are constants. The derivative can be obtained using 1.33 as follows

$$T(t + \delta) = a + b(t + \delta) \qquad (1.35a)$$

$$T(t - \delta) = a + b(t - \delta) \qquad (1.35b)$$

$$\text{Then} \quad T(t + \delta) - T(t - \delta) = 2b\delta \qquad (1.36)$$

Using equation 1.33

$$\frac{dT(t)}{dt} = \lim_{\delta \to 0} \frac{2b\delta}{2\delta} = b \qquad (1.37)$$

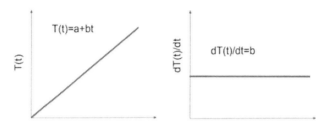

Figure 1.13: Time derivative of a linear function.

Example 1.2: Derivative of a quadratic function

A quadratic function is defined as

$$T(t) = a + b \cdot t^2 \tag{1.38}$$

Then

$$T(t + \delta) = a + b(t + \delta)^2 = a + b(t^2 + 2t\delta + \delta^2) \tag{1.39a}$$

$$T(t - \delta) = a + b(t - \delta)^2 = a + b(t^2 - 2t\delta + \delta^2) \tag{1.39b}$$

The derivative is

$$\frac{dT(t)}{dt} = \lim_{\delta \to 0} \frac{4bt\delta}{2\delta} = 2bt \tag{1.40}$$

Here are some important properties of the derivative of any function

$$\frac{d}{dt}(a \cdot T(t)) = a \cdot \frac{dT(t)}{dt} \tag{1.41}$$

Here a is a constant. Also

$$\frac{d}{dt}(T_1(t) + T_2(t)) = \frac{dT_1(t)}{dt} + \frac{dT_2(t)}{dt} \tag{1.42}$$

Here, $T_1(t)$ and $T_2(t)$ are two different continuous functions.

Table 1.1: Some functions and their derivatives

$T(t)$	$dT(t)/dt$	$T(t)$	$dT(t)/dt$
$e^{a \cdot t}$	$a \cdot e^{a \cdot t}$	$cos(a \cdot t)$	$-a \cdot sin(a \cdot t)$
t^a	$a \cdot t^{a-1}$	$sin(a \cdot t)$	$a \cdot cos(a \cdot t)$
$ln(a \cdot t)$	a/t	$tan(a \cdot t)$	$a/cos^2(a \cdot t)$

1C. Distance and speed

Distance, time and speed

Let us revisit the trip from Chiang Mai to Bangkok. During the trip you decided to run the following experiment. You have kept a track of the speed and recorded it every 20 minutes. You then started constructing a table similar to Table 1.2. By the time you arrived at the destination, you used the table to make a graph like figure 1.14. You asked yourself a question, can we calculate the total distance the car drove between the start and the destination using the recorded table and the figure obtained.

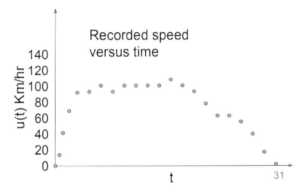

Figure 1.14: Graph of the recorded speed versus time.

Table 1.2: Some functions and their derivatives

Record index j	0	1	2	3 ......
Speed (km/hr)	0	10	42	70 ......

You recall that the speed you recorded has a unit of km/hr or you can put it in m/sec. It is a unit of a distance per time. Hence, speed by itself is a rate of change of distance with respect to time.

$$speed = u = \frac{\Delta x}{\Delta t} \qquad\qquad (1.43)$$

In your experiment, the time interval is constant and equals 20 minutes. The change of distance however depends on the speed at the time of the record or the record index, j, as in table 1.2. In terms of the record index, one can write the speed at a specific record as

$$speed_j = u_j = \frac{\Delta x_j}{\Delta t} \tag{1.44}$$

Based on equation 1.44 one could say that to know the change of distance at any record index j, we need to multiply the recorded speed by the time interval or

$$\Delta x_j = u_j \cdot \Delta t \tag{1.45}$$

Now, here is the problem. For equation 1.45 to be accurate we need to assume that the speed was constant during the whole-time interval Δt, which is 20 minutes or $1/3$ hours in our example.

Table 1.3: Speed and rate of change of distance.

Index j	0	1	2	....	n
$u_j (km/hr)$	0	10	42	....	u_n
$\Delta x_j (km)$	0	10/3	42/3	....	$u_n \cdot \Delta t$
$x_j (km)$	0	10/3	10/3+42/3	....	$\sum_{j=0}^{n} \Delta x_j$

This assumption is visualized in figure 1.15. Following this idea, one can construct table 1.6 for the distance covered at each record index. Notice that time interval $\Delta t = \frac{1}{3}$ hr. The third row in table 1.6 can be visualized as in figure 1.17 when assuming constant speed over the time interval. In this case the distance $\Delta x_j = u_j \cdot \Delta t$ is the area of a rectangle of height u_j and width t. The total distance covered after n time intervals is obtained by adding the multiplication of each measured speed by the time interval from $j = 0$ to n.

$$x_n = \sum_{j=0}^{n} u_j \cdot \Delta t \tag{1.46}$$

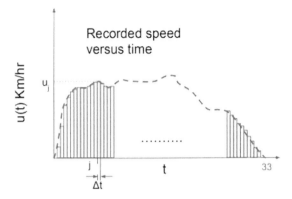

Figure 1.15: Visualization of the assumption of constant speed during each record time interval.

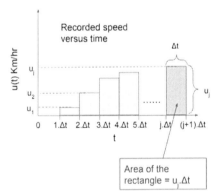

Figure 1.16: Visualization of the third column of table 1.6.

There are two points to mention here:

- First, each $\Delta x_j = u_j \cdot \Delta t$ multiplication is actually the area of the highlighted rectangle.
- Second, the total distance is the summation of these many areas. In total, it is the area under the speed-time curve.

At a record index j:
- Speed, u_j, is the range of change of distance over the time interval.
- Distance, Δx_j, is the area of a rectangle with height u_j and width Δt.
- The total distance, x_j, is the sum of the areas of all rectangles from index 0 to j.

Introducing integration

Of course, a smart observer would say, the speed was never constant during the 20 minutes time intervals. Hence, the measurements done before were error-some. We need much smaller intervals in order to be more confident assuming a constant speed during that time and hence to get closer to the proper solution. As a matter of fact, we need to reduce Δt so much such that it becomes infinitesimal. In this case we replace Δt by the infinitesimal interval dt.

In the earlier example, the time at each record index was $t_j = j \cdot \Delta t$. When infinitesimal time intervals are used, then the time can be assumed to be a continuous variable, t. The distance and speed at any index record, j, become functions of time. Mathematically, we can write this by taking the limit when $\Delta t \to 0$ which transfers the discrete representation of the displacement x into a continuous function of time as in figure 1.17.

$$\lim_{\Delta t \to 0} x_j = x(t) \tag{1.47a}$$

$$\lim_{\Delta t \to 0} u_j = u(t) \tag{1.47b}$$

Equation 1.46 will as well transfer to a continuous form as

$$x(t) = \lim_{\Delta t \to 0} \sum_{j=0}^{n} \Delta u_j \Delta t = \int_{t=0}^{t_{max}} u(t) dt \tag{1.48}$$

Here, t_{max} is the total trip time. When the intervals become small, the discrete speed measurement is replaced by the speed as a function of time. The summation is now replaced with another operator, Integration. From equation 1.48, the total distance can be measured directly by integrating the speed function over a time range starting from t=0 and ending with the total trip time.

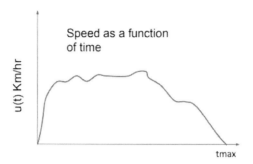

Figure 1.17: Speed as a continuous function of time.

Speed	Derivative of distance function with respect to time $u(t) = \frac{dx(t)}{dt}$	(1.49)
Distance	Integration of speed function over time $x(t) = \int u(t)dt$	(1.50)

Notice that at the beginning, when t = 0, the car is not moving and hence the speed is zero. Mathematically we would say, if the distance is constant, and equals xo for example, then the derivative of constant is zero.

$$u(t) = \frac{dx_o}{dt} = 0, \text{where } x_o \text{ is constant.} \qquad (1.51)$$

So, if distance was originally starting from a location x_o that is not zero, would equation 1.49 be still valid? We can examine that as follows. If the distance x was originally at x_o, then the displacement should take this into account. For instance, if we move to a location x_1, then the displacement is $x_1 - x_o$. As a function of time, we can write the displacement in a general form as $x(t) - x_o$. Hence, the speed as the derivative of the displacement can be written as

$$u(t) = \frac{d(x(t)-xo)}{dt} = \frac{dx(t)}{dt} - \frac{dx_o}{dt} = \frac{dx(t)}{dt} + 0 = \frac{dx(t)}{dt} \qquad (1.52)$$

Equation 1.52 gives us the same result as 1.49. If this is the case, one needs to correct equation 1.50 to include the presence of a constant in the result of the integration. One simple correction can be written as

$$x(t) - xo = \int_{t=0}^{tmax} u(t)dt, \text{ or} \qquad (1.53)$$

$$x(t) = xo + \int_{t=0}^{tmax} u(t)dt \tag{1.54}$$

We usually call such a constant, xo, as the initial distance.

Velocity

As we are preparing to end our experiment and feeling good about ourselves, the smart student showed some concern. He said: "Yes, we measured the distance from the speed but we did not measure the displacement vector we have discussed earlier" That really hurts. Yes, we overlooked the direction of the car when we measured the speed. We should have kept this in mind. Lucky enough, we had the GPS data handy, and we could correspond to every speed record with the direction at that moment.

When we combine the "amplitude" which is the speed and the direction, we obtain a vector that is commonly referred to as velocity. If we define a general displacement vector, $\bar{r}$, that has two components x and y as

$$\bar{r} = x\hat{x} + y\hat{y} = (x, y) \tag{1.55}$$

then the velocity vector of the car in figure 1.18 can be written as

$$\bar{u} = u_x\hat{x} + u_y\hat{y} = (u_x, u_y) \tag{1.56}$$

where

$$u_x = \frac{dx}{dt} \text{ and } u_y = \frac{dy}{dt} \tag{1.57}$$

In a vector form we can write the following equation

$$\bar{u} = \frac{d}{dt}(x, y) = \frac{d\bar{r}}{dt} \tag{1.58}$$

In this case the displacement vector can be written in a similar way as

$$\bar{r} = \bar{r}_o + \int_{t=0}^{tmax} \bar{u}(t)dt \tag{1.59}$$

Where $\bar{r}_o$ is the initial displacement vector.

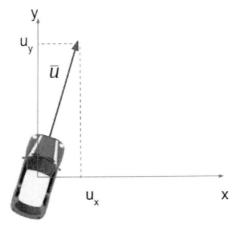

Figure 1.18: Velocity vector visualization

1D. Summary

- Displacement is a vector quantity that is described by an amplitude (length) and direction.
- A vector in two-dimensional space can be written as
  - $\bar{u} = (u_x, u_y) = u_x \hat{x} + u_y \hat{y}$
  - $\bar{u} = |\bar{u}|(cos\theta, sin\theta)$
    - $|\bar{u}| = \sqrt{u_x^2 + u_y^2}$
    - $\theta = tan^{-1}(u_y/y_x)$
- A vector in three-dimensional space can be written as
  - $\bar{u} = (u_x, u_y, u_z) = u_x \hat{x} + u_y \hat{y} + u + z\hat{z}$
- Two vectors summation and subtraction
  - $\bar{u}_1 + \bar{u}_2 = (u_{1,x} + u_{2,x}, u_{1,y} + u_{2,y}, u_{1,z} + u_{2,z})$
  - $\bar{u}_1 - \bar{u}_2 = (u_{1,x} - u_{2,x}, u_{1,y} - u_{2,y}, u_{1,z} - u_{2,z})$
- Two vectors dot product
  - $\bar{u}_1 \cdot \bar{u}_2 = u_{1,x}u_{2,x} + u_{1,y}u_{2,y} + u_{1,z}u_{2,z}$
- A variable is a parameter that changes and we perform our measurement depending on its change.
  - Variable: time, in the oven temperature example
  - Variable: distance, in the car speed recording example.
- A function represents the dependance of a quantity on a variable or more.
- Differentiation is the rate of change of a function with respect to a variable
  - $\frac{df(t)}{dt} = \lim_{\delta \to 0} \frac{f(t+\delta)-f(t-\delta)}{2\delta}$
- Integration is the area under the curve defined by the function in the y axis and the variable in the x axis.
  - $\int_{t=0}^{t_{max}} f(t)dt = \lim_{\Delta t \to 0} \sum_{j=0}^{n} f(t_j), n = \frac{t_{max}}{\Delta t} \ and \ t_j = j\Delta t$
- Speed is the rate of change of distance with time
  - $u(t) = \frac{dx(t)}{dt}$
- Distance is the integration of speed over time
  - $x(t) = x_o + \int u(t)dt, x_o$ is the initial distance.

- Velocity is the rate of change of the displacement vector
  - $\overline{u}(t) = \frac{d\overline{r}(t)}{dt} = \left(\frac{dx(t)}{dt}, \frac{dy(t)}{dt}, \frac{dz(t)}{dt} \right)$
- Displacement vector is the integration of velocity
  - $\overline{r}(t) = \overline{r}_o + \int \overline{u}(t)dt$, $\overline{r}_o$ is the initial displacement, or
    - $x(t) = x_o + \int u_x(t)dt \cdot y(t) = y_o + \int u_y(t)dt \cdot z(t) = z_o + \int u_z(t)dt$
- Scalar quantities:
  - Time, t.·Distance, x.·Speed, u.·Temperature, T.
- Vector quantities:
  - Displacement: $\overline{r}$.·Velocity, $\overline{u}$.
- Time has units of seconds or s.
- Displacement and distance have units of meter or m
- Velocity and speed have units of meter per second or m/s.

1E. Review questions

1. Select two cities in your country or region and write the displacement vector between them by filling the following table.

Staring city	Ending city	Distance (km)	Angle (°)	Displacemen t vector

2. Write which of the following quantities are scalar or vector

Displacement	Distance	Amplitude	Angle	Direction

3. Work out the numbers and calculate the displacement vector in the salesman example between Chiang Mai-Bangkok and Khon Kaen. Follow two approaches and compare the results.

   1. Direct calculation of displacement between Chiang Mai to Khon Kaen
   2. As a sum of two displacement vectors.

4. Using equations 1.7, prove that the amplitude of the vector $\overline{u}$ equals U. Note that the amplitudes is $|\overline{u}| = \sqrt{u_x^2 + u_y^2}$.

5. Prove equation 1.11.

6. Write the dot product of the following vectors:

   - $(1,-1) \cdot (1,1)$
   - $(1,0) \cdot (0,1)$
   - $(0.5,0.5) \cdot (1,1)$
   - $(0.4,0.2) \cdot (0.5,4)$

7. Write the derivative of the following functions:

   - $T(t) = 3 + 4t \cdot T(t) = 5 \cdot t^3$
   - $T(t) = 4cos(2t) \cdot T(t) = 3t + e^{5t}$

8.   Write the amplitude and phase of the following complex numbers

  • $z = 1 + i1$ • $z = i2$ • $z = 2 - i1$ • $z = -3 + i6$

9.   Perform the following experiment with a colleague and fill the table below:

  •   Select fixed time intervals (for example 10 seconds.
  •   Start walking with a fixed base and measure the distance over the intervals.
  •   Repeat the previous two steps but with a different time interval and walking base for five times.

Index j	Time interval Δt_j	Speed u_j	Distance Δx_j
0			
1			
2			
3			
4			

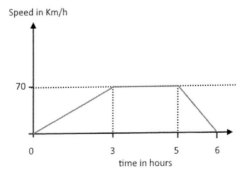

Figure 1.19. Speed versus time

10.   If the car speed as a function of time can be plotted as in figure 1.20, then use the definition of integration as the area under the curve to calculate the total trip distance from the figure. Hint, revisit the area of triangles and rectangles.

11.   For the speed graph in figure 1.20, if the car during the trip was always going in the north-east direction with an angle of $60o$ to the east axis then answer the following:

1.   Write the displacement vector of the trip.
2.   Write the velocity vector for the portion of the trip over which the car had a constant speed of 70 km/hr.
3.   Write an expression for the speed as a function of time for the first 3 hours of the trip.
4.   Write an expression of the speed as a function of time for the last hour of the tip.

CHAPTER TWO

MORE ON MOTION

2A. Expanding our knowledge of motion

Summary of quantities and units

The following two tables summarize some units we will use in this section as well as a reminder of some vector and scalar quantities that we covered in the last chapter.

Table 2.1: Some scalar quantities and their units.

Table 2.2 Some vector quantities and their units.

Quantity	Unit	Quantity	Unit
Distance	m	Displacement	m
Time	s	Velocity	m/s
Speed	m/s	Acceleration	m/s^2
Mass	kg	Force	Newton (N) kg m/s2

Acceleration

In the first chapter we stated that for a car that is moving in a certain direction, the motion can be represented by two vectors:

- Displacement vector, $\bar{r}$, that tells the distance and direction the car covers with reference to a starting point.
- Velocity vector, $\bar{u}$, that tells the speed and direction of the car movement at any point of space and instant of time.

Equation 1.58 in chapter one states the relation between the two vectors

$$\bar{u}(t) = \frac{d\bar{r}(t)}{dt} \tag{2.1}$$

That is the velocity at any point is the rate of change of the displacement vector. In terms of integration one can follow equation 1.54 to write the following equation.

$$\bar{r}(t) = \bar{r}_o + \int_0^t \bar{u}(t)dt \tag{2.2}$$

where $\bar{r}_o$ is the initial displacement vector. There is a third vector quantity that one needs to introduce which represents the change of speed with time. This quantity is called acceleration. Mathematically, we need to define the acceleration vector, $\bar{a}(t)$, as the rate of change of the velocity vector.

$$\bar{a}(t) = \frac{d\bar{u}(t)}{dt} \tag{2.3}$$

Acceleration then has units of m/s^2. To get a good understanding, let us think about a cannon shooting a metal ball as in figure 2.1. We can break the process into three stages:

1.   Before the explosion, the cannon ball is at rest.
2.   Right at the explosion, the cannonball started to increase its speed.
3.   When the explosion ends the cannonball moves with constant speed ignoring friction by the cannon interior wall.

We can visualize the motion by trying to plot the speed and acceleration of the cannon ball versus time.

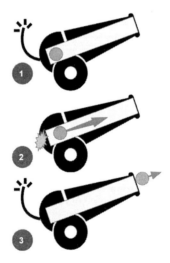

Figure 2.1: Stages of cannon ball movement.

1.   In the initial stage the cannonball is at rest, so it has zero speed, u_o. There is nothing moving it. So, the acceleration is zero as well, $|\bar{a}| = 0$.
2.   During the explosion time ($0 < t < \tau$) the ball starts to move with increasing speed. We can for simplicity assume that the rate of

change of speed is constant. In other words, the acceleration vector has a constant amplitude, $|\overline{a}| = a$, and a direction pointing at the same direction as the cannon.

3.   After the explosion finishes, there is nothing pushing the ball anymore. Hence, the speed remains constant. The acceleration is now zero back again, $|\overline{a}| = 0$.

To mathematically represent the cannonball motion, one can say that during the explosion time (from $t = 0$ to $t = \tau$) the acceleration has a constant amplitude, $|\overline{a}| = a$. We know from equation 2.3 that acceleration is the rate of change of velocity. In a similar way to the relation between displacement and velocity, we can write the following integration relation between velocity and acceleration vectors.

$$\overline{u}(t) = \overline{u}_o + \int_0^t \overline{a}(t)dt \qquad (2.4)$$

Here, $\overline{u}_o$ is the initial velocity. In our case both velocity and acceleration are pointing in the same direction as the cannon. Hence, we can only be concerned with the scalar values of speed and acceleration amplitudes. Finally, the acceleration amplitude is assumed to be constant during the explosion, $|\overline{a}| = a$. Using these facts, we can rewrite equation 2.4 as follows.

$$u(t) = \int_0^t a \cdot dt = a \cdot t \qquad (2.5)$$

Equation 2.5 reads that during the explosion period, $0 < t < \tau$, the speed is increasing linearly versus time with a slope of a m/s^2. When the explosion ends, the acceleration is now back to zero, as there is nothing trying to move the ball further. The speed however will remain constant at the maximum value it reached, $a \cdot \tau$. We can write this motion as

$$u(t) = \begin{cases} a \cdot t & 0 < t < \tau \\ a \cdot \tau & t > \tau \end{cases} \qquad (2.6)$$

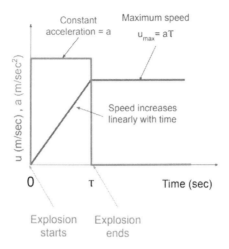

Figure 2.2. Plots of speed and acceleration of the cannon ball inside the cannon.

The acceleration is speed per time, hence the acceleration unit is (m/s) / s = m/s².

Gravitational acceleration

One obvious example of acceleration would be gravity. Most of us are familiar with the gravitational acceleration $g = 9.81 m/s^2$. If we let an apple fall from a height h, its speed increases linearly till it hits the ground. What happens there? Recall from the cannonball example, the explosion causes a push, and that push causes the cannonball to accelerate. So, whenever there is a push or a pull there should be an acceleration. When we let the apple fall from a height, gravity pulls the apple down. Once, there is a pull then there is an acceleration. Luckily, this one was measured a long time ago and has been confirmed. It is known as the gravitational acceleration.

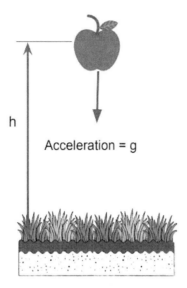

Figure 2.3: An apple falling to the ground due to gravity with an acceleration that equals g.

Gravitational acceleration is different from planet to planet. For example, the moon is smaller than Earth and hence its gravitational acceleration is one sixth of that on Earth. The acceleration is constant in the case of gravity and it is pointing downwards towards the center of Earth. If we consider y axis to point upwards, then

$$\overline{a} = -g\hat{y} \tag{2.7}$$

The negative sign in equation 2.7 means that the acceleration is pointing to the $-y$ direction or downwards.

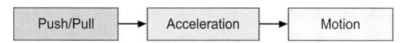

Figure 2.4: Push and pull effect on motion diagram.

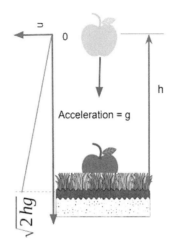

Figure 2.5: Velocity and displacement of a free-falling object due to gravity.

We know from the cannonball example that when the acceleration is constant then the velocity will be linearly increasing

$$\bar{u}(t) = -\int_0^t g dt \hat{y} = -g \cdot t \hat{y} \qquad (2.8)$$

The negative sign of the velocity again indicates downwards motion. The displacement vector is the integration of the velocity vector.

$$\bar{r}(t) = \bar{r}_o - \int_0^t g \cdot t dt = \left(h - \frac{1}{2} g t^2 \right) \qquad (2.9)$$

The initial displacement vector is $\bar{r}_o = h\hat{y}$. When the object reaches the ground, its displacement vector should be 0.

$$h - \frac{1}{2} g t_{max}^2 = 0 \rightarrow t_{max} = \sqrt{\frac{2h}{g}} \qquad (2.10)$$

t_{max} is the total time it takes for the object to free fall to the ground under the effect of the gravitational acceleration. Using equation 2.8, the maximum velocity is

$$u_{max} = g \cdot t_{max} = \sqrt{2gh} \qquad (2.11)$$

2B. Projectiles

In the previous section, we discussed the motion of a cannonball inside a cannon till it reaches the tip of the cannon. We did not however discuss what happens when it leaves the cannon?

Projectile calculation

As we discussed earlier, the cannonball gains acceleration during the explosion time then it sustains a constant speed along the direction of the cannon. Once it leaves the cannon gravity will be constantly pulling the cannonball towards the earth till it finally manages to bring it down. The path the cannonball takes from exiting the cannon till it reaches the ground is called the projectile.

Let us assume that we put the cannon such that its tip is just at the ground level. It is also positioned at an angle θ with respect to the ground, or the x-axis. We put the origin of the x-y coordinate system right at the tip of the cannon as shown in figure 2.6. As in the figure, in this problem the initial velocity is the velocity of the cannonball right at the tip of the cannon. Remember this is the maximum velocity in the case study in the previous section. We can express it in the following vector form.

$$\bar{u}_o = u_{x,o}\hat{x} + u_{y,o}\hat{y} = |\bar{u}_o|(\cos\theta\,\hat{x} + \sin\theta\,\hat{y}) \tag{2.12}$$

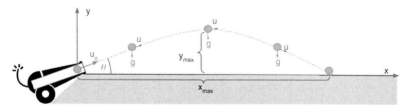

Figure 2.6: Projectile motion for the cannonball assuming the cannon tip is touching the ground.

The acceleration is pointing in the negative y direction

$$\bar{a} = a_x\hat{x} + a_y\hat{y} = 0\hat{x} - g\hat{y} = -g\hat{y} \tag{2.13}$$

The velocity is the integration of acceleration

$$\bar{u}(t) = \bar{u}_o + \int_0^t \bar{a}(t)dt \tag{2.14a}$$

$$= |\bar{u}_o|(\cos\theta\,\hat{x} + \sin\theta\,\hat{y}) - gt\hat{y} \qquad\qquad (2.14b)$$

$$= |\bar{u}_o|cos\theta\hat{x} + (|\bar{u}_o|sin\theta - gt)\hat{y} \qquad\qquad (2.14c)$$

The displacement vector is the integration of velocity

$$\bar{r}(t) = \bar{r}_o + \int_0^t \bar{u}(t)dt \qquad\qquad (2.15a)$$

$$= |\bar{u}_o|cos\theta \cdot t\hat{x} + \left(|\bar{u}_o|sin\theta \cdot t - \frac{1}{2}gt^2\right)\hat{y} \qquad\qquad (2.15b)$$

Equation 2.15b above describes the progress of the location of the canon ball as a function of time. This progress has two main characteristic points, one of which is when the canon ball is at the maximum height and the other is the distance at which it touches the ground.

The characteristic points on the projectile path

1. The point at which the cannonball touches the ground. At this point, the y value reaches zero again. To find the maximum distance, we need to put the component in the displacement vector in equation 2.15b to be zero and find the value of x that gives this solution.
2. The point at which the cannonball reaches a maximum height. At this point, the velocity is along the x axis only. Hence the x component of the velocity should be zero. To find the maximum height then we need to set the y component of velocity in equation 2.14c to zero and find the value of x that gives this solution. This value of x is then used to find y_{max}.

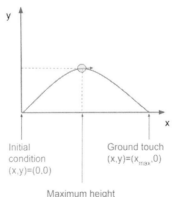

Figure 2.7: The projectile path as height versus distance and the main characteristic points.

To get the maximum distance the projectile takes to reach the ground, we find the time, t_{max}, it takes such that the y component of the displacement vector goes back to zero.

$$|\bar{u}_o| \sin \theta \cdot t_{max} - \frac{1}{2} g t_{max}^2 = 0 \rightarrow t_{max} = \frac{2|\bar{u}_o| \sin \theta}{g} \qquad (2.16)$$

The maximum distance is then obtained by substituting t_{max} into the x component of equation 2.15b.

$$x_{max} = |\bar{u}_o| \cos \theta \cdot t_{max} = \frac{2|\bar{u}_o|^2 \cos \theta \sin \theta}{g} \qquad (2.17)$$

So, what is $\cos \theta \sin \theta$? Let us recall the expressions in 1.16a and 1.16b in the first chapter and use them to find a simplified form of the multiplication.

$$\sin \theta \cos \theta = \frac{1}{2i} \left(e^{i\theta} - e^{-i\theta} \right) \frac{1}{2} \left(e^{i\theta} + e^{-i\theta} \right) \qquad (2.18a)$$

$$= \frac{1}{2} \cdot \frac{1}{2i} \left(e^{i2\theta} - 1 + 1 - e^{-i2\theta} \right) \qquad (2.18b)$$

$$= \frac{1}{2} \cdot \frac{1}{2i} \left(e^{i2\theta} - e^{-i2\theta} \right) \qquad (2.18c)$$

$$= \frac{1}{2} \cdot \sin(2\theta) \qquad (2.18d)$$

Using the expression in 2.18d in equation 2.17, the maximum distance is

$$x_{max} = \frac{u_o^2 \sin(2\theta)}{g} \qquad (2.19)$$

For simplicity we set the amplitude of the initial velocity, or initial speed, as $|\bar{u}_o| = u_o$.

The maximum height is achieved when the velocity has only a component in x. That is when the y components in equation 2.14c equals zero. Hence, we calculate the time, t_y, at which the height is maximum as

$$u_o \sin \theta - g t_y = 0 \rightarrow t_y = \frac{u_o \sin \theta}{g} \qquad (2.20)$$

Using this time into equation 2.15b

$$y_{max} = u_o \sin\theta \cdot \left(u_o \frac{\sin\theta}{g}\right) - \frac{1}{2}g\left(u_o \frac{\sin\theta}{g}\right)^2 \qquad (2.21a)$$

$$= \frac{u_o^2 sin^2\theta}{g} - \frac{u_o^2 sin^2\theta}{2g} = \frac{u_o^2 sin^2\theta}{2g} \qquad (2.21b)$$

Projectile with:
Initial speed of u_o ·Initial angle θ
Has:
Maximum distance: $x_{max} = \dfrac{u_o^2 \sin(2\theta)}{g}$
Maximum height: $y_{max} = \dfrac{u_o^2 sin^2\theta}{2g}$

Now let us prove that the maximum distance is achieved for $\theta = 45^o$. We know that $x_{max} = \dfrac{u_o^2 \sin(2\theta)}{g}$. Maximum value of $sin(2\theta)$ is 1. That is when $2\theta = 90^o$ or $\theta = 45^o$.

Example 2.1

What is the speed needed to make the cannonball complete one turn around the earth?

First, let us set the initial angle to 45°. The maximum distance is then, $x_{max} = \dfrac{u_o^2}{g}$. Set x_{max} to equal the earth circumstance or $x_{max} = 40075\ km$. The initial speed needed then is

$$u_o = \sqrt{x_{max} \cdot g} = \sqrt{4007500 \cdot 9.81} = 6270\ km/s$$

Figure 2.8: A cannon shooting a ball for a complete round around the earth.

How long would it take to complete the whole round?

$$t = 2u_o \cdot sin\theta = 2 \cdot 6270.05 \cdot sin(45^o) = 904s.$$

That is 904/60 = 15 minutes. So, it will take the ball 15 minutes to complete one round around the Earth.

2C. Circular motion

At this point, we know that the push, by the cannon explosion for example, and the pull of gravity, cause the object to accelerate away (in case of push) or towards (in case of gravity). The question is what other forms of motion can cause acceleration? There is and we most certainly have experienced it on many occasions. We experience it when the car makes a roundabout turn or a u-turn. At that moment, we feel that something is pushing us away from the center of the car's rotation. This is known as circular motion.

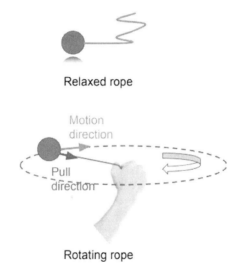

Relaxed rope

Rotating rope

Figure 2.9: An object rotating in a circular motion experiences a pull towards the center.

Circular motion Description

Let us perform the following experiment illustrated in figure 2.9.

- First, tie a ball to a rope.
- Observe that, when the ball is laying on the ground, the rope is relaxed.
- Hold the end of the rope and start to rotate in a circular motion
- Observe that the rope is now clearly stretched.

This stretch tells us the following:

- There is a pull of the ball towards the center.
- This pull is there to sustain the rotation.
- Without this pull the ball would fly away.

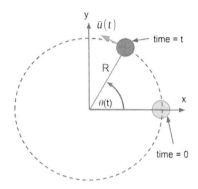

Figure 2.10: Circular motion of an object around the center.

Can we prove this observation mathematically? Let us first have a top view of the rotating rope and set a coordinate system that has an origin at the center of rotation where the hand is located as illustrated in figure 2.10. What we know about the rope is that it has a length of R meters. We could as well measure that it makes several rotations, v, per one second. That means the time of one rotation is $1/v$ seconds. For every complete rotation, the angle $\theta(t)$ completes 360 degrees or 2π. Hence, at any time t the angle measured is

$$\theta(t) = \frac{2\pi t}{\text{time of rotation}} \qquad (2.22a)$$

$$= \frac{2\pi t}{1/v} = 2\pi v t \qquad (2.22b)$$

We define $\Omega = 2\pi v$, where Ω is commonly referred to as the angular frequency and it has units of radian/second or rad/s in short. As in figure 2.10, at any time interval t the object displacement vector can be written as.

$$\bar{r}(t) = x(t)\hat{x} + y(t)\hat{y} \qquad (2.23)$$

Where

$$x(t) = R \cdot cos(\theta(t)) = R \cdot cos(\Omega t) \qquad (2.24a)$$

$$x(t) = R \cdot sin(\theta(t)) = R \cdot cos(\Omega t) \qquad (2.24b)$$

From Chapter One, we know that the velocity is the rate of change of the displacement vector.

$$\overline{u}(t) = \frac{d\overline{r}(t)}{dt} = \frac{dx(t)}{dt}\hat{x} + \frac{dy(t)}{dt}\hat{y} \qquad (2.25)$$

Using the relations in table 1.2, we can write the following

$$\frac{dx(t)}{dt} = -R\Omega \cdot sin(\Omega t) \qquad (2.26a)$$

$$\frac{dy(t)}{dt} = R\Omega \cdot cos(\Omega t) \qquad (2.26b)$$

The velocity vector is then

$$\overline{u}(t) = R\Omega(-sin(\Omega t)\hat{x} + cos(\Omega t)\hat{y}) \qquad (2.27)$$

The velocity vector, in the first quarter, is pointing to the negative x direction and positive y direction as shown by the arrow in figure 2.10. The speed of the rotation, or the amplitude of the velocity vector, is

$$|\overline{u}(t)| = \sqrt{u_x^2(t) + u_y^2(t)} = \sqrt{R^2\Omega^2 sin^2(\Omega t) + R^2\Omega^2 cos^2(\Omega t)} = R\Omega$$
$$(2.28)$$

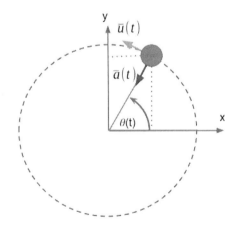

Figure 2.11: Circular motion of an object around the center.

Rotational acceleration

Acceleration vector is the rate of change of the velocity vector.

$$\overline{a}(t) = \frac{d\overline{u}(t)}{dt} = \frac{du_x(t)}{dt}\hat{x} + \frac{du_y(t)}{dt}\hat{y} \qquad (2.29)$$

$$\frac{du_x(t)}{dt} = -R\Omega^2 \cdot cos(\Omega t) \qquad (2.30a)$$

$$\frac{du_y(t)}{dt} = -R\Omega^2 \cdot sin(\Omega t) \qquad (2.30b)$$

The amplitude of the acceleration vector is

$$|\overline{a}(t)| = R\Omega^2 \qquad (2.31)$$

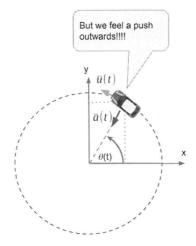

Figure 2.12: A car moving continuously in a circular motion.

The acceleration vector is pointing inwards towards the center of rotation. If a car kept a continuous rotation in a roundabout, what do we as passengers feel? Based on the earlier discussion in figure 2.11 and equations 2.30, one expects to feel a pull towards the center along the direction of acceleration. However, in reality we feel the opposite. We might have already experienced it. When the car turns in a roundabout or a sharp U-turn, we feel a continuous push away from the center of rotation. This seems to contradict our last observation where the acceleration is pointing inwards towards the center.

In this chapter, we will work this out based on logic and common sense. However, later we will find a proper statement to explain this phenomenon.

Stating Newton's third law of motion

Consider the following example: Ali is standing inside a truck that is at rest. Hassan the reckless driver, without a warning, pressed the gas paddle suddenly.

Question: What would Ali experience?

- A. A push towards the rear of the truck, opposite to the direction of the movement.
- B. A push towards the front of the truck along the direction of the movement.

We might answer this question quickly, it is A. Ali feels a push towards the rear of the truck. That is opposite to the direction of acceleration. This common-sense answer is a well-known law, Newton's third law of motion.

"For every action, there is an opposite and equal reaction."

This is exactly why we feel the push away from the center of rotation in a direction opposite to that of the acceleration. Remember the acceleration is due to the car being pulled towards the center not to move away from the circular motion. However, we inside that rotating car we react to this pull by an opposite push away from the center of rotation.

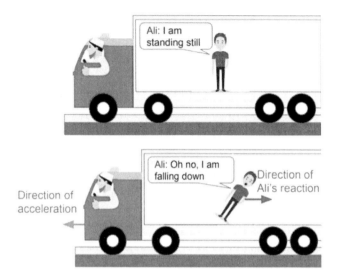

Figure 2.13: A truck standing still while a passenger, Ali, is standing. Alis does not feel any push or pull as long as the car is not moving. Once the car starts moving, Ali experiences a push in a direction opposite to that of the car movement.

Example 2.2: Artificial gravity

Can we use this basic concept to think of a way that allows astronauts in space to feel a gravity like that on Earth? The question might sound like a sci-fi movie type of thing. Though, by now you have learnt enough to make a good guess. Let us use the following hints:

1. Gravity is felt by an acceleration constant g = 9.81 m/s^2.

2. Continuous rotation causes an acceleration pointing inwards towards the center of rotation.

3. For a person inside a rotating vehicle, he experiences a reaction opposite to the acceleration and with equal effect.

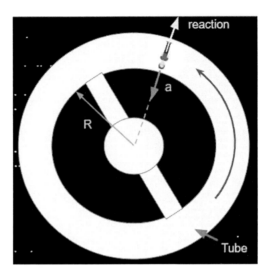

Figure 2.14: A spaceship rotating around the center to generate an acceleration similar to gravity.

So, what we need to do is to design a spaceship with a rotating tube around the center. To feel gravity, we need to make sure that the amplitude of the generated acceleration equals g.

$$|\bar{a}| = R\Omega^2 = g \rightarrow \Omega = \sqrt{g/R} \qquad (2.32)$$

If the spaceship engineers designed a tube with a radius R = 10 m, then how many rotations per seconds are needed to generate gravity?

$$\Omega = \frac{9.81}{10} = 0.981 \ rad/s \qquad (2.33)$$

In terms of rotation, the spacecraft needs to rotate by

$$v = \frac{\Omega}{2\pi} = \frac{0.981}{2\pi} = 0.157 \frac{rotation}{s} \qquad (2.34a)$$

$$= 0.157 \cdot 60 = 9.5 \ rpm \qquad (2.34b)$$

Hence, the tube needs to rotate by 9.5 rotations per minute to sustain an acceleration push of 9.81 m/s² towards the outer side of the tube. People will then walk with their heads pointing to the center of the craft.

2D. Summary

- Acceleration is the rate of change of velocity: $\overline{a}(t) = \frac{d\overline{u}(t)}{dt}$.
- Velocity is the integration of acceleration over time: $\overline{u}(t) = \overline{u}_o + \int_0^t \overline{a}(t)dt$.
- Acceleration has units of meter per square second or m/s^2.
- Gravity has a constant acceleration:
  - Amplitude g = 9.81 m/s^2
  - Direction downwards towards the center of Earth ($-y$ for example).
- An object that was initially at rest falls down with
  - An acceleration of: $\quad\quad \overline{a}(t) = -g\hat{y}$.
  - A velocity of: $\quad\quad\quad \overline{u}(t) = -g \cdot t\hat{y}$.
  - A displacement vector: $\overline{r}(t) = \left(h - \frac{1}{2}g \cdot t^2\right)\hat{y}$., h is the original height of the object.
- The time the object takes to reach the ground (when r=0) is: $t_{max} = \sqrt{\frac{2h}{g}}$.
- When there is an acceleration, it means that the object is being pushed forward/upward or pulled backward/downwards.
  - A cannonball experiences acceleration during the explosion period that pushes it forwards.
  - A cannonball has no more acceleration after the explosion and hence it moves with constant speed.
- When exiting the cannon, the cannon ball has:
  - An initial displacement: $\overline{r}_o = (0,0)$.
  - an initial velocity: $\overline{u}_o = u_o(cos\theta, sin\theta)$
    - u_o is the initial speed $u_o = |\overline{u}_o|$
    - θ is the angle the cannon makes with respect to the ground.
  - An acceleration due to gravity: $\overline{a} = -g\hat{y}$
- The velocity of the cannonball is: $\overline{u}(t) = \overline{u}_o - g \cdot t\hat{y}$
  - $u_x = u_o cos\theta$
  - $u_y = u_o sin\theta - g \cdot t$

- The displacement of the cannonball is: $\overline{r}(t) = \overline{u}_o \cdot t - \frac{1}{2}g \cdot t^2 \hat{y}$
  - $x = u_o t \cdot cos\theta$
  - $y = u_o t \cdot sin\theta - \frac{1}{2}g \cdot t^2$
- This type of motion is commonly referred to as projectile motion
- The time the cannonball takes to reach the ground is obtained when y=0
  - $t_{max} = \frac{2u_o sin\theta}{g}$
- The distance it covers (along x-axis) till it touches the ground is
- $x_{max} = \frac{2u_o^2 cos\theta sin\theta}{g} = \frac{u_o^2 sin(2\theta)}{g}$
- The maximum height of the ball is achieved when $uy = 0$
  - Time to maximum height: $t_y = \frac{u_o sin\theta}{g}$
  - Maximum height: $y_{max} = \frac{u_o^2 sin^2(\theta)}{2g}$
- An object that rotates on a circular trajectory of radius R, it has
  - Displacement vector: $\overline{r}(t) = R \cdot (cos\theta(t), sin\theta(t))$
    - $\theta(t)$ is the angle a line connects the center to the object makes with respect to the x-axis at time t.
  - $\theta(t) = 2\pi vt.$
    - v is the number of rotations the object makes per second. v has units of 1/second or 1/s
      - It is common to use rotations per minute (rpm)
      - $rpm = v \cdot 60$
    - $\theta(t) = \Omega t$
      - Angular frequency: $\Omega = 2\pi v$
      - Angular frequency has units rad/s
- Velocity: $\overline{u}(t) = R\Omega \cdot (-sin(\Omega t), cos(\Omega t))$
  - Amplitude: $R\Omega$
  - Direction: Along $\hat{\theta}$.
- Acceleration: $\overline{a}(t) = R\Omega^2 \cdot (-cos(\Omega t), sin(\Omega t))$
  - Amplitude: $R\Omega^2$
  - Direction: Along $-\hat{r}$

2E. Review questions

1. A child is holding a toy car on a perfectly smooth surface. He started moving it with constant acceleration of 2 m/s² for 5 seconds then he let it go. Write an expression for the toy car velocity versus time and its displacement vector. Assume that the child was pushing the car along the x axis direction.

2. A cannon has a length of 2 meters and it is aimed at an angle of 30 degrees upwards from the ground. If the time the ball takes from firing the explosion in the cannon till reaching the tip of the cannon is 1 second. Calculate the constant acceleration knowing that the explosion time lasts for 0.2 seconds.

3. An apple is left to free fall from a distance of 4 meters above the ground. Calculate the time it takes to reach the ground and the maximum speed it reaches just before hitting the ground.

4. If the same apple in the previous question was on the surface of the moon, what would be the total time and maximum speed when it reaches the ground?

5. In a football game, your friend is standing 5 meters away from you and he is waving to you so you can pass him the ball. Select a proper initial speed and initial angle so that you guarantee that the ball reaches him.

6. Prove that $\frac{y_{max}}{x_{max}} = \frac{1}{4} tan\theta$. Hint, use the expression of $x_{max} = \frac{2u_o^2 sin\theta cos\theta}{g}$ and $y_{max} = \frac{u_o^2 sin^2(\theta)}{2g}$.

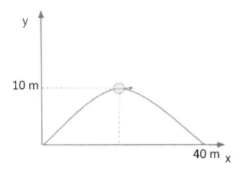

Figure 2.15: The projectile path for problem 7.

7.   The graph in figure 2.15 shows the projectile path. From the numbers present in the graph, estimate the initial speed and initial angle. Hint: use the expression driven in question 6.

8.   Prove that when the projectile is at $y = y_{max}$ then $x = \frac{x_{max}}{2}$.

9.   A monkey is standing on a 2-meter-high banana tree that is 4 meters away from you. You decide to throw him a banana. What would be the needed initial angle and initial speed for the banana to reach the monkey.

10.   A metallic ball is tied to a 1-meter rope and a motor starts to rotate it with a speed of 20 rounds/s. Calculate the speed and rotational acceleration that results from the motion.

11.   Prove that $cos^2\theta + sin^2\theta = 1$. Hint, use the relations in equation 1.16 in chapter One.

12.   If a spaceship is rotating at 2 rpm, what is the radius of the ship that is needed so the astronauts could feel earth's gravity on the ship?

13.   A car entered a roundabout at a speed of 40 km/hr and started rotating with the same speed. If the roundabout radius is 5 meters, what is the push acceleration that the passengers feel inside the car?

CHAPTER THREE

NEWTON'S FIRST LAW OF MOTION

3A. First law and newton's relativity

Newton's first law of motion states the following

> "In an inertial frame of reference, an object either remains at rest or continues to move at a constant velocity, unless acted upon by a force."

One can derive a fast understanding that when we are in a frame of reference an object does not move if it was not originally moving or continues to move at the same speed that it was originally moving with unless an acting force is applied on this object. We might follow the second part as nothing would change unless a force is applied. But what does it mean by a frame of reference?

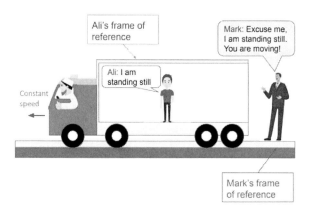

Figure 3.1: Hassan's truck is moving at a constant speed. Ali feels no push or pull at all. He thinks that he is standing still. On the platform Mark observes that he is the one standing still but Ali is moving with the truck. Both are correct because they look at things from their own frames of reference.

Frame of reference

Let us revisit Ali and Hassan truck again. Ali's frame of reference is where he finds himself at rest, no acceleration due to a pull or push. In this case, it is the truck that moves at a constant speed (acceleration is zero) as illustrated in figure 3.1. Mark on the other hand is standing on a platform on the road. For Mark, it is the road's platform where he finds himself at rest. In this case, the platform is Mark's frame of reference. From Mark's point of view, Ali's frame of reference is moving forward. From Ali's point of view, it is Mark's platform that is moving backwards. In each frame of reference one can set a system of coordinates that seems to be fixed around a specific point we call it the origin.

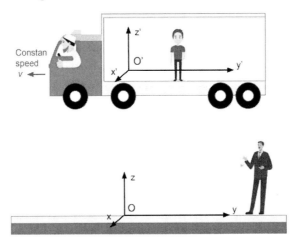

Figure 3.2: Setting up two coordinate systems, one for each frame of reference. Ali sets coordinate system with an origin located at some corner inside the truck. Mark on the other hand sets a coordinate system with an origin point placed at some point on the road's platform.

Ali can define the location of objects in his frame of reference (the moving truck) using the coordinate system. Mark can do the same for his frame of reference (The road's platform) and set the coordinate system as illustrated in figure 3.2. Each one of these coordinate systems has its own point of origin where displacements are defined respectively as shown in the figure.

The race from Mark's frame of reference

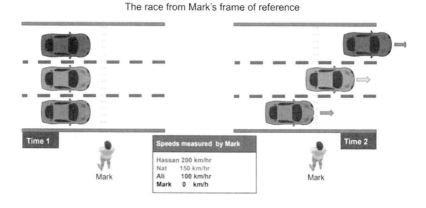

The race from Nat's frame of reference

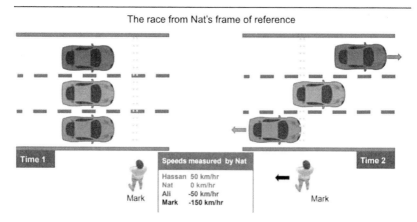

Figure 3.3: Visualization of the race at two different times, time 1 and time 2. The top snapshot is taken from Mark's frame of reference. Mark is still and not moving in the snapshot. However, the three cars have changed their position in time 2. The lower snapshot shows the same race at the same times time 1 and time 2 but from Nat's frame of reference. Nat's car seems to be at the same position in the two pictures however Hassan seems to be moving forward while Ali's car and Mark seem to move backwards.

Relative speed

Being a reckless driver, Hassan thought he would be better fit for racing. For some reason Ali thought of the same thing, though he is an extremely careful driver. Their friend Nat, who happens to be a reasonable driver, decided to join the race. Mark on the other hand decided to watch them from the sideline. We wanted to visualize this race at two-time snaps. To make things more exciting we decided to take those snaps from too different frames of reference:

1. Mark's frame of reference standing on the sideline.
2. Nat's frame of reference being inside a moving car with a medium speed.

These snaps are shown in figure 3.3. Looking at the snapshots and the speeds measured by both Mark and Nat, we can summarize the observations as follows:

1. From Mark's point of view, the three cars were moving along the positive x-axis with speeds: u_1, u_2 and u_3. Mark himself was not moving then his speed was zero ($u_4 = 0$).
2. From Nat's point of view things looked different:
   a. Nat himself was not moving.
   $$u_2' = u_2 - u_2 = 0 \qquad (3.1)$$
   b. Mark was the fastest and he moves backward with speed:
   $$u_4' = 0 - u_4 = -u_4 \qquad (3.2)$$
   c. Ali was moving backward as well but with a slower speed:
   $$u_3' = u_3 - u_2 \qquad (3.3)$$
   d. Hassan was the only one moving forwards with a slow speed of:
   $$u_1' = u_1 - u_2 \qquad (3.4)$$

Relative velocity

In the previous example we talked about speed forward and backward. We have already familiarized ourselves with the velocity vector in the previous chapters. If we assume that the forward direction matches the positive x-axis, then the quantities in figure 3.3 can be re-written in vector form as in table 3.1.

Table 3.1: Velocities calculated from two frame references, Mark standing on the side and Nat moving with his car at a speed of 150 km/hr along the positive x-axis direction.

Velocity (km/hr)	From Mark's frame	From Nat's frame
Hassan's	$\bar{u}_1 = 200\hat{x}$	$\bar{u}'_1 = 200\hat{x} - 150\hat{x} = 50\hat{x}$
Nat's	$\bar{u}_2 = 150\hat{x}$	$\bar{u}'_2 = 150\hat{x} - 150\hat{x} = 0\hat{x}$
Ali's	$\bar{u}_3 = 100\hat{x}$	$\bar{u}'_3 = 100\hat{x} - 150\hat{x} = -50\hat{x}$
Mark's	$\bar{u}_4 = 0\hat{x}$	$\bar{u}'_4 = 0\hat{x} - 150\hat{x} = -150\hat{x}$

What if Ali and Hassan were driving in opposite directions keeping the same speed values as in table 3.1? What would Hassan's speed be from Ali's frame of reference? From Ali's frame of reference, he sees Hassan approaching him with a speed $\bar{u}'_1 = -200\hat{x} - 100\hat{x} = -300\hat{x}$

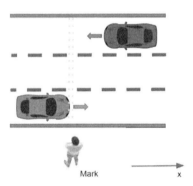

Figure 3.4: Visualization of two cars driving at opposite directions from Mark's frame of reference.

Table 3.2: Ali and Hassan's relative velocities from both Mark's and Ali's frame of references.

	Mark's frame (km/hr)	Ali's frame (km/hr)
Ali	$\overline{u}_3 = 100\hat{x}$	$\overline{u}_3' = 0\hat{x}$
Hassan	$\overline{u}_1 = -200\hat{x}$	$\overline{u}_1' = \overline{u}_1 - \overline{u}_3$
		$= -200\hat{x} - 100\hat{x}$
		$= -300\hat{x}$

What if Hassan was coming from an intersection, how would Ali see his velocity? In this situation, as in the upper graph in figure 3.4 Hassan is moving along the negative y direction while Ali is moving along the positive x direction. From Ali's point of view, Hassan is approaching him with a speed with an angle of $tan^{-1}(200/100) = 63^o$.

$$|\overline{u}'| = \sqrt{u_x'^2 + u_y'^2} = \sqrt{100^2 + 200^2} = 244 km/hr. \quad (3.5)$$

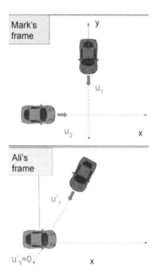

Figure 3.5: Ali and Hassan's car movements from two different frames of reference.

Table 3.3: Relative velocity between Ali and Hassan's cars when approaching an intersection from two frames of reference: Mark standing and Ali's car moving along the positive x direction.

	Mark's frame (km/hr)	Ali's frame (km/hr)
Ali	$\bar{u}_3 = 100\hat{x}$	$\bar{u}_3' = 0\hat{x}$
Hassan	$\bar{u}_1 = -200\hat{y}$	$\bar{u}_1' = \bar{u}_1 - \bar{u}_3$
		$= -200\hat{y} - 100\hat{x}$

3B. Inertia

Let us recall the statement of Newton's first law of motion: "In an inertial frame of reference, an object either remains at rest or continues to move at a constant velocity, unless acted upon by a force." In the previous section, we explained the meaning of frame of reference. However, we did not talk about the term "inertial".

The definition of inertia

If you look up many physics textbook, you would find the following definition of inertia

> "the resistance of any physical object to any change in its velocity."

We understand from the previous section that velocity is a vector which has two main characteristics: amplitude (or speed) and direction. Hence, the definition above can be restated as the resistance of any object to the change of its speed or the direction of motion. One of the situations that fits under changing speed is to move an object that was originally at rest.

Let us imagine the following scenario returning to Ali and Hassan truck in figure 3.1. If Mark, who was on the platform, built a truck model out of papers. Now, Hassan stopped his truck right next to Mark's model. They looked identical except that we know Mark's model is made of paper while the original truck is made of heavy metallic parts. Ali and Mark decided to try to change the speed of both vehicles and compare their experiences. We could right away realize without a doubt that the paper truck is much lighter than the original truck. Moving, changing speed or direction of the lighter object is much easier than a heavier object.

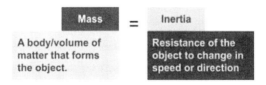

Figure 3.6: Mass and inertia are the same quantity of the object defined from two different points of view.

We use lighter and heavier objects in our daily life to deal with weight. Don't we? Or if you read in scientific books, you would be familiar with the word mass.

So, is inertia related to weight and mass? Let us examine what mass is. Mass deals with what material fills the object's volume. So, a plastic ball is a spherical volume filled with plastic. If we fill the same volume with gas, it will have a lower mass. If we fill the volume with iron, it will have a larger mass. It is easier to move, change speed or direction of the plastic ball compared to the iron ball. Hence, an object with a less mass will have less inertia (or less resistance to movement) and vice versa. One can then say that Inertia and mass are correlated. As a matter of fact, both inertia and mass are the same quantity defined from two different perspectives. Hence, the unit of inertia is as well Kg.

Mass and gravity

We know that gravity is felt through a pull with a constant acceleration g towards the ground (or the center of Earth). In figure 3.8 both the apple and the bowling ball will fall with the same acceleration and reach the ground at the same time. So, what is the difference? Where is the mass or inertia playing a role here?

The difference could be easily observed by the marks that both objects leave when they hit the ground.

- The bowling ball leaves a more pronounced and obvious mark on the ground.
- The apple on the other hand would leave a less obvious mark and it could even break itself. So, why would this be?

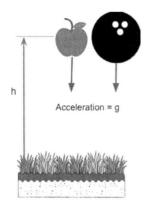

Figure 3.7: An apple and a bowling ball in free fall.

3C. Force

Let us recall again the statement of Newton's first law of motion: "In an inertial frame of reference, an object either remains at rest or continues to move at a constant velocity, unless acted upon by a force." In the previous sections, we discussed frame of reference and inertia. However, we did not mention the last word in the statement, "force."

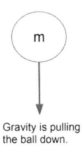

Gravity is pulling
the ball down.

Figure 3.8: Gravitational pull of an object of mass equals m.

Gravitational force

For the example in figure 3.7, we know that gravitational acceleration is the same for both objects. Hence both the apple and the bowling ball will change their velocity similarly. However, we know that the bowling ball, being more massive, will have larger inertia or larger resistance to the change in velocity. Hence, we expect that gravity will need to provide stronger pull to the bowling ball compared to the apple. This action of pulling (or pushing) is called force. In our example this pulling is called the gravitational force. Force is measured in the units of Newton or in short N.

From the definitions we reached, the force exerted by gravity on an object depends on its mass as well as the gravitational acceleration.

$$\overline{F}_g = -mg\hat{y} \qquad\qquad (3.6)$$

In equation 3.5 the y-axis is assumed to be pointing upwards; hence the gravitational pull is pointing downwards or at the negative y direction. Looking at the force equation one can say that the force has units of $N = Kg \cdot m/s^2$. We commonly call the amplitude of this gravitational force weight.

A proper unit of weight is then N not kg as we commonly do. However, we typically use (*weight*/*g*) in place of weight. A ball with a mass *m*, will experience a gravitational force pulling it down, or $weight = m \cdot g$. If we place this ball on the moon, it will experience less gravitational acceleration to pull it down.

$$g_{Moon} = \frac{1}{6}g \rightarrow weight_{Moon} = \frac{1}{6}weight_{Earth} \qquad (3.7)$$

The weight of the ball on the moon = 1/6 weight on Earth.

> Notice that the mass of the object is not affected by the place it is in. The ball has the same mass on Earth and on the moon. This is because the mass depends on the matter that forms the object. That does not change by the location in this context.

Springs and weight measurement

One of the common ways to measure weight is using a spring as shown in figure 3.9. Most of us have experienced a spring inside a broken pen, broken toy or other devices. It is basically a wire that is curled to the form in the figure. This structure gives the spring an elastic property. So, what does this spring do? When you press on the spring it is squeezed and you feel a push on your hand exerted by the spring. The more the spring is squeezed, the stronger the force it exerts on the hand.

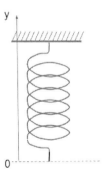

Figure 3.9: A spring formed by curling a wire and hanging it on the wall to be used for weight measurement.

When the spring is pulled it is stretched and you feel a pullback force exerted by the spring on your hand. The more the spring is stretched (below breaking point), the stronger the force it exerts on the hand. This is illustrated in figure 3.10.

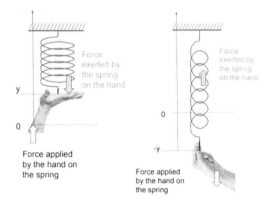

Figure 3.10: (Left) a hand is pushing the spring upwards, and it is experiencing a force that is pushing the hand downwards. (Right) A hand is pulling the spring downwards and it is experiencing a force that pulls the spring upwards.

Hooke's law

If we apply a force (pull or push), the spring reacts in the opposite direction. The strength of the force depends on the amount of pressing or stretching. The pressing is represented by a positive amount of displacement y of the tip of the spring from the original rest position. Here, we assumed that the spring axis is along the y direction. When the spring is pressed the displacement is assumed to be positive as in the figure. However, stretching is represented by a negative displacement value. The spring force reacts in the opposite direction. Hence, it is defined as

$$\overline{F}_s = -K \cdot y \, \hat{y} \tag{3.8}$$

The constant K is defined as the spring constant with units of N/m.

Example 3.1: Using spring as a scale

The negative sign in Hooke's law indicates the spring force is opposite to the spring displacement. When a load is placed at the end of the spring, the spring is stretched by y in the negative direction. The spring stretches until the weight equates the spring force (or the spring force comprises the gravitational force).

$$|\overline{F}_s| = |\overline{F}_g| \tag{3.9}$$

$$K \cdot y = m \cdot g \rightarrow y = \frac{m \cdot g}{K} \tag{3.10}$$

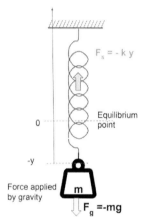

Figure 3.11: Using the spring as a scale.

We define a quantity $S = \frac{g}{K}$, which has units of m/Kg.

$$y = S \cdot m \tag{3.11}$$

Calibration: So, to make a working scale, we need to perform a process called calibration. For that, we load several well-known masses: $m_1, m_2, \ldots, m_n$. Then we measure the corresponding displacements $y_1, y_2, \ldots, y_n$ and extract a linear relation between displacement and mass.

$$y_j = S \cdot m_j, j = 1,2,3,\ldots,n \tag{3.12}$$

Hence, we can mark a corresponding weight for each displacement next to the spring.

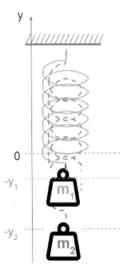

Figure 3.12: Calibration of the spring to make a scale.

- By now we understand that inertia and mass represent the same quantity but looking from two different points of view (units of Kg).
- The objects' resistance to movement or change of velocity depends on their mass.
- Force in the form of pulling or pushing causes the object to change their velocity (start moving, stop moving, change speed or direction.)
- Without applying force, objects continue to move with no change in velocity or do not move if they are originally at rest.
- Then force exists during the period of changing the velocity. (Assuming no change in mass)

3D. Summary

- Frame of reference:
  - A set of coordinates used to define locations and velocities.
  - Objects moving with different constant velocities will have different frames of references.
- Relative speed:
  - Scalar quantity
  - Objects move in opposite or the same direction
  - $u_2' = u_2 - u_1$
    - Speed of object 2 relative to object 1
    - u_2' is the speed of object 2 from object 1 frame of reference
    - If object 2 is in the opposite direction to object 1, then
      - u_2 is negative
      - u_2' has a larger negative value
    - If object 2 is in the same direction as object 1, then
      - u_2 is positive
      - u_2' has a smaller positive value, 0 or a negative value
- Relative velocity:
  - Vector quantity
  - $\overline{u}_2' = \overline{u}_2 - \overline{u}_1$
    - $u_{2,x}' = u_{2,x} - u_{1,x}$
    - $u_{2,y}' = u_{2,y} - u_{1,y}$
    - $u_{2,z}' = u_{2,z} - u_{1,z}$
- Inertia:
  - The resistance of the object to change its velocity Units of Kg
- Mass:
  - Deals with what the object is made of. Units of Kg
- Inertia = Mass
- Force:
  - The action of pulling or pushing an object Units of Newton or N
  - Units of $Kg \cdot m/s^2$
  - Gravitational force:
    - $\overline{F}_g = -m \cdot g\hat{y}$
    - Weight is the gravitational force and has units of N not Kg.

- Springs exert force when pressed or stretched.
  - At equilibrium
    - Displacement = 0
    - Force = 0
  - Pressed:
    - Displacement = y
    - Force=$-Ky$
  - Stretched:
    - Displacement = -y
    - Force = Ky
- Hooke's law:
  - $\overline{F}_s = -K \cdot y\hat{y}$
- Holding a mass m, the spring reaches rest when
  - $|\overline{F}_s| = |\overline{F}_g|$
  - $y = -\frac{m \cdot g}{K}$

3E. Review questions

1. Ann and Nat are running on a track and both are going in the same direction. Ann is running at 20 km/hr while Nat is running at 25 km/hr. What is Ann's relative speed from Nat's frame of reference?

2. Complete the following table

Car 1 velocity $\bar{u}_1$ (km/hr)	Car 2 velocity $\bar{u}_2$ (km/hr)	Relative velocity $\bar{u}_2 - \bar{u}_1$ (km/hr)
$20\hat{x} + 40\hat{y}$	$-20\hat{x} + 40\hat{y}$	
$-50\hat{x} + 50\hat{y}$		$100\hat{x} - 100\hat{y}$
	$80\hat{y}$	$30\hat{x}$
$10\hat{x} + 50\hat{y}$		$-10\hat{x} - 50\hat{y}$

3. For the two cars in figure 3.13 write the relative velocity of the red car and blue cars from the following frames of references A person standing on the road. The red (light) car. The blue (dark) car. The red car is moving at a speed of 100 km/hr and the blue car is moving at a speed of 120 km/hr at an angle of 60o to the x axis. Hint: use the relation in equations 1.7a and 1.7b to represent the velocity of the blue car.

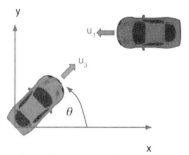

Figure 3.13 Two cars moving at different speeds and angles.

4.   Underline the objects with higher inertia and write down the time it
      takes to reach the ground from a free fall.

Object 1	Object 2	Height (m)	Time to ground (s)
Football	Table tennis ball	10	
Car toy	Electric car	5	
Watermelon	Lemon	15	
A book	A tablet	2	

5.   A metallic ball that has a mass of 5 kg. Calculate the weight of this
      ball measured at the following locations: [a] Earth [b] Moon [c]
      Mercury [d] Venus [e] Mars

6.   A spring that has a constant of 200 N/m is used to measure the weight
      of small languages in the airport. When you hang your handbag on it
      you can measure a displacement of 10 cm downwards. What is the
      weight and mass of the bag?

7.   You performed an experiment to measure the spring constant for an
      old spring that you found in the storage. You simply placed known
      masses and measured the displacement it causes. Doing so, you
      obtained the data in the table to the side. Can you estimate the spring
      constant based on these data? Hint: Remember that in any experiment
      there is always an error produced by several factors. So, when
      determining the spring constant, you might need the average of the
      results obtained from each measurement.

Mass (kg)	Displacement (m)
0	0
5	5.1
10	9.9
15	14.8
20	20.2

CHAPTER FOUR

NEWTON'S SECOND LAW OF MOTION

4A. Momentum

The rise of momentum

Let us revisit the cannon example. During the explosion time, there is a force applied on the cannonball that makes it gain speed. After the explosion finishes, or $t > \tau$, that force vanishes, and the ball only experiences gravitational force pulling it downwards. However, in the forward direction the ball moves with a constant speed as in figure 4.1. Regardless of the ball mass, the projectile will remain the same (maximum distance and height.) However, from observations we know that a plastic ball would make much less impact when hitting the ground compared to metallic cannonball. The impact of the projectile depends on the mass of the object as well as on its speed. That brings forward a new quantity which is momentum. The momentum vector is defined as the multiplication of mass and velocity.

$$\overline{P} = m \cdot \overline{u} \qquad\qquad\qquad (4.1)$$

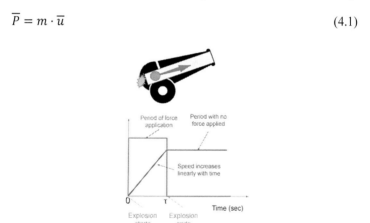

Figure 4.1: The forward motion of the cannonball inside the cannon

So, we can say that when the cannon has an explosion, a force exists, and the ball starts to gain "momentum" along the cannon direction. When the explosion effect is finished, the pushing force becomes zero keeping only the gravitational force pulling it down. Hence during being in the air, the ball will sustain a constant forward, P_x, momentum component while the vertical component, P_y, keeps reducing due to gravity. Once it reaches the ground, the ball speed will start to change. Hence, a force is exerted on the ground. This force causes an impact on the ground. The cannonball loses its speed and hence momentum. Here, we neglect bouncing off ground and assume that the ball becomes completely at rest once it hits the ground. This behavior is depicted in figure 4.2.

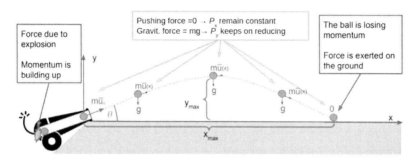

Figure 4.2: Momentum and force of a cannonball on a projectile.

From the illustration in figure 4.2 one can state that when force exists there is a change in momentum. In other words, if a constant force is applied on the object, its momentum will keep increasing. Once the force is removed, the momentum will remain constant. We can say the force is the rate of change of the momentum or

$$\overline{F} = \frac{d\overline{P}}{dt} \tag{4.2}$$

Using equation 4.1 in equation 4.2

$$\overline{F} = \frac{d(m \cdot \overline{u})}{dt} \tag{4.3}$$

If the mass of the object (say the cannonball) remains constant (not broken into pieces for example) then we can derive the following

$$\overline{F} = m\frac{d\overline{u}}{dt} = m \cdot \overline{a} \tag{4.4}$$

Force is then the mass times acceleration. Recall that when we studied gravitational force (or weight) in the previous section we wrote the force as $\overline{F}_g = -m \cdot g\hat{y}$. The gravitational acceleration is $\overline{a} = -g\hat{y}$. Hence the gravitational force is as well $\overline{F}_g = m \cdot \overline{a}$, when $\overline{a} = -g\hat{y}$.

4B. Force and acceleration

The second law's statement

The results above bring us directly to Newton's second law of motion:

"The acceleration of an object is directly related to the net force and inversely related to its mass"

On the light of this law, let us study the difference in the motion of two objects with different masses when equal constant force $(\overline{F}_o\hat{x})$ is applied on each as in figure 4.3. When we kick both balls with the same exact force, would we naturally expect them to accelerate similarly? To answer this let us check the force equation 4.4.

$$\overline{F} = m \cdot \overline{a} \rightarrow \overline{a} = \frac{\overline{F}}{m} \qquad\qquad (4.5)$$

Plastic golf ball with mass = m,   Metallic golf ball with mass = m$_2$

Figure 4.3: Equivalent forces are applied to two different gold balls with different masses. Which one would gain a larger acceleration?

In our example, the two balls will have different acceleration which are inversely proportional to their masses, $\overline{a}_1 = \frac{F_o}{m_1}\hat{x}$ and $\overline{a}_2 = \frac{F_o}{m_2}\hat{x}$ respectively. Knowing that the metallic ball is heavier than the plastic one, $m_2 > m_1$, then it should be accelerated less than the lighter ball. This is clear from the equations above and by intuition.

Example 4.1

For the example in figure 4.3, assume a force of 1 N was applied on both balls. If the plastic ball weighs 100 gram and the metallic ball is 500 gram and the kicking time lasts 100 ms, what is the speed and momentum that each ball gains?

$$\bar{a}_1 = \frac{1N}{0.1kg}\hat{x} = 10\,\hat{x}\ m/s^2$$

$$\bar{a}_1 = \frac{1N}{0.5kg}\hat{x} = 2\,\hat{x}\ m/s^2$$

$\bar{u}_1 = \bar{a}_1 \cdot t = 10t\,\hat{x}\ m/s$, when t=100 ms=0.1 s, then $\bar{u}_1 = 10 \cdot 0.1\hat{x} = 1\,\hat{x}\ m/s$

$\bar{u}_2 = \bar{a}_2 \cdot t = 2t\hat{x}\ m/s$, when t=100 ms=0.1 s, then $\bar{u}_2 = 2 \cdot 0.1\,\hat{x} = 0.2\,\hat{x}\ m/s$

The momentum each ball receives is the multiplication of speed by mass

$$\bar{P}_1 = m_1 \cdot \bar{u}_1 = 0.1 \cdot 1\hat{x} = 0.1\hat{x}\ kgm/s$$

$$\bar{P}_2 = m_2 \cdot \bar{u}_2 = 0.5 \cdot 0.2\,\hat{x} = 0.1\,\hat{x}\ kgm/s$$

4C. Friction

From example 4.1, we could conclude that both golf balls will move with constant speeds u_1 and u_2 indefinitely. In other words, there is nothing there that is telling us when they will stop. This contradicts our common sense and what we observe daily. If we kick a ball, it will move then it will start reducing speed till it eventually comes to a complete stop. Is there anything wrong with Newton motion? The answer is within this range of speed (very small compared to the speed of light) there is nothing wrong in Newton's equation of motion. We just did not include one important factor in motion which is loss. In our situation, the loss comes from a quantity known as friction. At the first glance, friction would sound to depend mainly on the surface roughness (neglecting air and wend effect). However, if we think carefully, the light ball will seem to logically move faster on a rough surface compared to the heavier one.

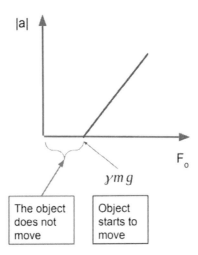

Figure 4.4: Amplitude of acceleration as a function of applied force when friction is present.

Friction force

So, it seems to logically think of friction as a force that resists movement, and it depends on both surface roughness (a factor to be introduced) and the weight of the object. The friction force as well must be opposite to the direction of motion. In other words, it must be slowing down the motion. If the object is moving along the positive direction of the x-axis, then the friction force can be written as

$$\overline{F}_{friction} = -\gamma mg\,\hat{x} \tag{4.6}$$

The constant γ is the friction coefficient, which is a unit-less quantity that depends on the material of both the surface and the object. For example, a rubber tire has less resistance, or small γ, when it moves on an icy road compared to asphalt. Also, a rubber tire will have smaller γ on asphalt compared to a tire made of rocks on the same asphalt road.

So, to move an object made of a certain material on a surface of another material, one needs to overcome the friction between these two materials. In other words, the applied force needs to overcome the friction force for the object to start moving as in figure 4.4. Applying larger force, $F_o\hat{x}$, results in acceleration along the x-axis can be written as

$$\overline{F} = F_o\hat{x} - \gamma mg\hat{x} = m \cdot \overline{a} \tag{4.7}$$

Then

$$\overline{a} = \left(\frac{F_o}{m} - \gamma g\right)\hat{x} \tag{4.8}$$

One can immediately realize that to move an object (obtain a positive acceleration) the applied force has to overcome friction or $F_o > \gamma mg$. Notice that, for lower value of force,

$F_o < \gamma mg$, there will be no negative acceleration, it remains zero and hence, we should correct equation 4.8 by stating clearly that it is valid for $F_o \geq \gamma mg$. If friction becomes negligible, such as the case in space, the slightest force should be able to move a massive object, and the object will keep moving indefinitely. On Earth or other planet surfaces, continuous force is needed to overcome friction to keep an object moving on the surface.

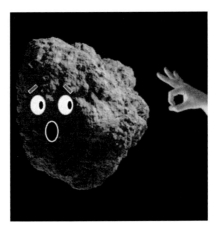

Figure 4.5: Space has almost no friction. A slight force applied on any object would cause an almost indefinite change in the object's motion.

For instance, to keep a car moving at a constant speed one needs to keep pressing the speed paddle. Pressing the speed paddle means applying a force, and we know force should result in acceleration. Acceleration would mean that speed should be increasing. However, we keep pressing the paddle only to keep the speed constant. The solution of this dilemma is the fact that the extra force applied by pressing the speed paddle is only to compensate for the loss resulted by the friction force between the tires and the road. This force is constantly slowing down the car.

Motion with friction

When an object moves under friction, the object starts to accelerate once the applied force is larger than the friction force as in equation 4.8. If the constant force is applied for a time of τ, then the velocity obtained right at time τ is

$$\bar{u} = \left(\frac{F_0}{m} - \gamma g\right)\tau\,\hat{x} \tag{4.9}$$

After time τ the force is removed and hence the object is de-accelerated.

$$\bar{F} = -\gamma m g\,\hat{x} = m \cdot \bar{a} \rightarrow \bar{a} = -\gamma g\,\hat{x} \tag{4.10}$$

The velocity is the integration of acceleration,

$$\bar{u} = \bar{u}_o + \int_0^t \bar{a}(t)dt \qquad (4.11a)$$

$$= \int_0^\tau \left(\frac{F_o}{m} - \gamma g\right) dt \,\hat{x} - \int_\tau^t \gamma g dt \,\hat{x} \qquad (4.11b)$$

$$= \left(\frac{F_o}{m} - \gamma g\right)\tau \,\hat{x} - \gamma g(t - \tau)\,\hat{x} \qquad (4.11c)$$

The equation can be simplified as

$$\bar{u} = \left(\frac{F_o}{m}\tau - \gamma g t\right)\hat{x} \qquad (4.12)$$

When the object stops at time t_{max}, its velocity has a zero amplitude

$$0 = \frac{F_o}{m}\tau - \gamma g t_{max} \qquad (4.13)$$

Hence

$$t_{max} = \frac{F_o \tau}{\gamma m g} \qquad (4.14)$$

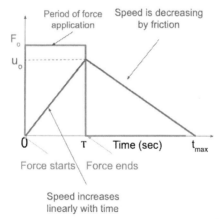

Figure 4.6: Visualization of an object motion under friction. For a constant force application of a time duration τ, the object's speed increases linearly. After lifting the force, the object reduces its speed in a linear fashion until it reaches a complete stop at time t_{max}.

These relations are visualized in figure 4.6. The figure shows a linear increase in speed until the time τ. Once the constant force F_o is lifted, the speed starts to decrease linearly, and the object stops at a time that equals

t_{max}. The value of this time as in equation 4.14 depends on the applied force, the time direction of the force application, the mass of the object and most importantly on the friction coefficient. So, the same object under the same force condition will have different t_{max} for different surfaces. Table 4.1 shows some values of the friction coefficients between different materials. The time the object takes from the moment the force is lifted till it reaches a complete stop, $t_{stop} = t_{max} - \tau$, is commonly known as the time to stop.

Table 4.1: Friction coefficients between different materials [2].

Materials	Friction
Rubber on concrete (dry)	0.68
Rubber on asphalt (dry)	0.68
Rubber on ice	0.15
Wood on wood	0.3
Steel on wood	0.57
Copper on steel	0.36

Time and distance to stop

While driving a car on a road, you decide to stop pressing on the speed paddle and measure the time it takes for the car to stop. This time is "the time to stop" for this particular car (mass and tire material) on this particular road (road material). From equation 4.14 and the definition of the time to sop, we can write the following expression

$$t_{stop} = \frac{F_0\tau}{\gamma mg} - \tau = \frac{\left(\frac{F_0}{m}-\gamma g\right)\tau}{\gamma g} \tag{4.15}$$

Notice that the nomination in equation 3.15 is the speed of the object right at the end of the force application. For the example in figure 4.7, this would be the time when the driver stopped pressing the speed paddle. We can define this speed as the amplitude of an initial velocity $\bar{u}_o = \left(\frac{F_0}{m} - \gamma g\right)\tau\,\hat{x}$. The time to stop can be written in terms of the speed as

$$t_{stop} = \frac{|\bar{u}_o|}{\gamma g} \tag{4.16}$$

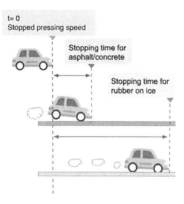

Figure 4.7: Visualized comparison of the time and distance to stop for a car driving on asphalt and ice.

The time to stop then depends on the initial speed, $|\overline{u}_o|$, of the car and the friction coefficient between the car and the road material. The initial speed is what you read in the car's speed meter when you first decided to stop pressing the speed paddle. Let us set the time origin to be the instant when we decided to stop pressing the speed paddle, which is as well the instant at which the force is lifted in figure 4.6. This is the time t=τ in the figure. In other words, we could simply be setting τ=0 in the new time coordinate. Using this definition, the velocity becomes

$$\overline{u}(t) = \overline{u}_o - \gamma g t \,\hat{x} \tag{4.17}$$

The displacement vector is obtained by integrating the velocity in equation 4.17 when setting the initial displacement to be zero.

$$\overline{r}(t) = \overline{u}_o \cdot t - \frac{1}{2}\gamma g t^2 \,\hat{x} \tag{4.18}$$

When the car stops, the time it takes to stop, t_{stop}, is as in equation 4.16. Using that in equation 4.18, we can define the distance to stop, r_{stop}

$$r_{stop} = \frac{|\overline{u}_o|^2}{\gamma g} - \frac{1}{2}\gamma g \left(\frac{|\overline{u}_o|}{\gamma g}\right)^2 = \frac{|\overline{u}_o|^2}{2\gamma g} \tag{4.19}$$

4D. Summary

- Momentum:
  - A vector quantity
  - The object's mass times its' velocity
    - $\overline{P} = m \cdot \overline{u}$
  - Presence of force indicates change in the momentum
    - $\overline{F} = \frac{d\overline{P}}{dt}$
  - When mass is constant
    - $\overline{F} = m \cdot \overline{a}$
- Force:
  - The object's mass times its acceleration
  - Acceleration depends on the object's mass and the force applied
    - $\overline{a} = \frac{\overline{F}}{m}$
  - Applying constant force
    - Heavier object accelerates slower.
- Friction:
  - A force that resists movement of one object over another.
  - On a flat surface
    - Amplitude depends on
      - Weight: mg
      - The roughness between the object and the surface
        - Friction coefficient γ
    - Direction
      - Opposite of the direction of motion
  - For an object of mass moving along the positive x-direction
    - $\overline{F}_{friction} = -\gamma mg \,\hat{x}$
  - Applying a force $F_o x$ on the object originally at rest
    - $F_o \leq \gamma mg$
      - Acceleration: $\overline{a} = 0$
      - Velocity: $\quad \overline{u} = 0$

- $F_o > \gamma mg$
  - Acceleration: $\overline{a} = \left(\frac{F_o}{m} - \gamma g\right) \hat{x}$
  - Velocity: $\overline{u} = \left(\frac{F_o}{m} - \gamma g\right) t \, \hat{x}$
  - Displacement: $\overline{r} = \frac{1}{2}\left(\frac{F_o}{m} - \gamma g\right) t^2 \, \hat{x}$
- For an object with initial velocity $\overline{u}_o = u_o x$
  - $F_o = 0$
    - Acceleration: $\overline{a} = -\gamma g \, \hat{x}$
    - Velocity: $\overline{u} = (u_o - \gamma g t)$
    - Displacement: $\overline{r} = \left(u_o t - \frac{1}{2}\gamma g t^2\right)$
    - Time it takes the object to stop
      - $\overline{u} = 0 \hat{x} \rightarrow t_{stop} = \frac{u_o}{\gamma g}$
    - Distance it takes the object to stop
      - $r_{stop} = \frac{u_o^2}{2\gamma g}$

4E. Review questions

1. Ann and Nat are running on a track and both are going in the same direction. Ann is running at 20 km/hr while Nat is running at 25 km/hr. If both Nat and Ann have the same momentum and Ann's mass is 60 kg, what is Nat's mass?

2. Complete the following table. Hint, remember that the momentum is in $kg \cdot m/s$ and the velocity is given in km/hr.

Car velocity $\bar{u}_{(}km/hr)$	Car mass $m(kg)$	Car momentum $\bar{P} = m \cdot \bar{u}$ $(kg \cdot m/s)$
$20\hat{x} + 40\hat{y}$	200	
$-50\hat{x} + 50\hat{y}$		$-417\hat{x} + 417\hat{y}$
	150	$83\hat{x}$
$10\hat{x} + 50\hat{y}$		$-1278\hat{x} = 6389\hat{y}$

3. For the example in figure 4.3, if the two golf balls are identical and they both have a mass of 100 g, if the forces applied are F1 = 1 N and F2 = 5 N respectively, what are the accelerations they gain? If the force remains for 100 ms, what are the speed and momentum each ball gains by the end of the force? Comment on the momentum in this question compared to example 4.1.

4. For the example in figure 4.2, calculate the momentum vector at three locations: [a] Right at the tip of the cannon. [b] At the location of maximum height. [c] When it reaches the ground. Note that the ball is 10 kg, the initial angle is 30° and the initial speed is 20 m/s.

5. Two objects are free falling from a latitude of 10 meters to the ground. The objects have masses 200 g and 1 kg respectively. What are the forces the objects exert on the ground at the moment of impact. Also, what is the momentum of each object at that moment?

6. You designed a rocket model that has a mass of 2 kg and was curious what would be the minimum force needed so that the rocket could start moving upwards?

7.   For the rocket in question 4c-3, A constant force of 40 N was applied
      to it for 2 seconds. What is the maximum height that the rocket could
      reach? — Hint: Divide the problem into two parts:

-    In the first part there are two acceleration components $\bar{a}_{up} = F/m\hat{y}$ point upwards and $\bar{a}_{down} = -g\hat{y}$ that points downwards. The total acceleration is the summation of these two accelerations. Use the relation between acceleration, velocity and displacement to calculate the speed and height at the end of the two seconds.

-    In the second part the acceleration is only pointing downwards $\bar{a}_{down} = -g\hat{y}$, but there is an initial speed (from the first part) and initial height (from the first part). Use the relations from the projectiles section or the relations between acceleration, velocity and displacement to know the maximum height.

8.   Use the values in table 4.1 to calculate the time and distance to stop
      for the following objects in table 4.2. All objects are moving in the
      positive x-direction.

Table 4.2. Time and distance to stop for different objects with different conditions.

Object	Surface	Object mass (kg)	Initial speed (km/hr)	Time to stop (s)	Distance to stop (m)
Rubber tire of a car	Asphalt	1400	50		
Rubber tire of a car	Ice	1400	50		
Steel container	Steel	1200	20		
Wooden tree log	Wood	60	15		

9.   Two cars are racing. The first car has a mass of 1500 kg while the second car has a mass of 1200 kg. Both cars accelerated at a constant amplitude of 8. 9 $m/s2$ for a time duration of 20 seconds. After that, both racers stopped pressing the speed paddle. Note that the road is asphalt.

1.   Which car would take a longer time to stop.
2.   Calculate the stop time and stop distance for each car.
3.   What would be the time to stop and distance to stop for both cars if it was raining on the day of the race if we are told that the friction coefficient of rubber on asphalt reduces to 0.4 when it is wet.

CHAPTER FIVE

NEWTON'S THIRD LAW OF MOTION

5A. Third law and free body diagram

Newton's third law of motion states that

For every action in nature there is an equal and opposite reaction.

One can always read this law as follows: If a force is applied by an object A on an object B, then object B will exert an equal but opposite force on object A. For example, object A is placed on a ground (object B) as shown in figure 5.1. Then gravitational force is applied on the ground by the object due to its mass.

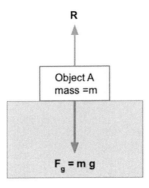

Figure 5.1: The ground reaction, R, to the gravitational force, F_g, applied by the object.

$$\overline{F}_g = -mg\hat{y} \tag{5.1}$$

The ground reacts with an equal but opposite force R.

$$\overline{R} = -\overline{F}_g = mg\hat{y} \tag{5.2}$$

The total force then exerted on the object is the summation of the two forces.

$$\overline{F}_{total} = \overline{F}_g + \overline{R} = 0 \tag{5.3}$$

The total force is zero, hence there is no acceleration, and the object remains still with no motion as expected.

Free body diagram

Let us revisit the friction problem we discussed in unit four. When a force in the x direction is applied on the object, a resisting friction force exists. What is the source of this force? We know that gravity is pushing the object down and the surface is reacting to this force upwards. Hence, this reaction of the surface to gravity causes the felt resistance of the object to motion with an amount that depends on the materials that form both surfaces. Hence, the amplitude of the friction force can better be written as.

$$|\overline{F}_{friction}| = \gamma|\overline{R}| \tag{5.4}$$

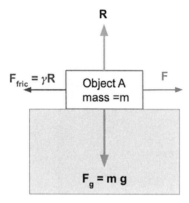

Figure 5.2: The free body diagram of an object moving under friction om a flat surface.

Figure 5.2 shows four arrows that represent the forces that are applied on the object. These forces are:

1. Gravitation force, $\overline{F}_g = -mg\hat{y}$, that pushes the object downwards.
2. Reaction force, $\overline{R} = -\overline{F}_g$, that pushes the object upwards.
3. Pulling force, $\overline{F}$, that is pulling the object in the positive x-direction.
4. Friction force, $\overline{F}_{friction} = -\gamma|\overline{R}|\hat{x}$, that pulls the object in the negative x-direction.

The diagram in figure 5.2 that shows the relative magnitudes and directions of the applied forces on the object is commonly referred to as the free body diagram. The total force exerted on the object is then

$$\overline{F}_{total} = \overline{F}_g + \overline{R} + \overline{F} + \overline{F}_{friction} \qquad (5.5a)$$

$$= (mg - mg)\hat{y} + (F - \gamma mg)\hat{x} = (F - \gamma mg)\hat{x} \qquad (5.5b)$$

In equations 5.5, we simply use the notation that $F = |\overline{F}|$. One important correction to equation 5.5b is that if the amplitude of the force is less than the friction force, then the total force must equal 0. Otherwise, it will indicate that object should be moving in the negative x-direction if $F < \gamma mg$, which is not true. So, we can put this correction as follows.

$$\overline{F}_{total} = \begin{cases} 0 & F \leq \gamma mg \\ (F - \gamma mg) & F > \gamma mg \end{cases} \qquad (5.6)$$

Now, equation 5.6 matches the intuition we have that the object would only move once the applied force overcomes friction. That is the reason we keep pressing on the speed paddle in the car to keep it moving at a constant speed. We are basically applying enough force to overcome friction between the rubber tire and the road.

5B. Motion on Tilted surfaces

The concept of object motion under friction with an applied force should be clear by now for a flat surface where the object lays horizontally. What about a tilted surface? From our daily observation, we noticed that when an object is placed on a flat surface and we start to tilt it, the object might remain still until we reach a specific angle after which the object starts to move downwards. The object's acceleration increases as we increase the tilt.

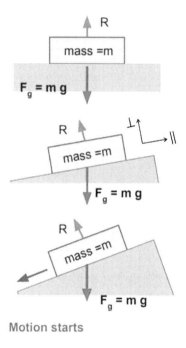

Figure 5.3: Observing an object motion under friction and different tilt angles.

Can we explain this observation based on what we have learned so far? The answer is yes, and all we need to do is to examine the free body diagram of the object under the different tilting conditions. The total force should be proportional to the object acceleration as we have learned from Newton's second law of motion.

When dealing with tilted surfaces, it is always a good practice to define two directions, one parallel to the surface and one perpendicular to it.

- ⊥ represents a direction normal to the surface that is pointing outwards.
- ∥ represents a direction parallel to the surface that is pointing upwards.

In a tilted surface, the reaction force is normal to the surface and its value equals the portion of the gravitational force that is normal to the surface,

$$\overline{R} = mg \cos \theta \; \hat{\perp} \tag{5.7}$$

The force that pulls the object down should be the portion of the gravitational force that is parallel to the surface and pointing downwards (to the left side in our case in figure 5.4.)

$$\overline{F} = -mg \sin \theta \; \hat{\parallel} \tag{5.8}$$

Friction force is proportional to the surface reaction as mentioned in equation 5.4 and it is pointing at a direction opposite to the direction of the intended motion. In this case, the friction must be pointing parallel to the surface in the upwards direction (to the right side).

$$\overline{F}_{friction} = \gamma |\overline{R}| \; \hat{\parallel} = \gamma mg \cos \theta \; \hat{\parallel} \tag{5.9}$$

The total force applied on the object is then

$$\overline{F}_{total} = \overline{F}_g + \overline{R} + \overline{F}_{friction} \tag{5.10a}$$

$$= \left(-mg \cos \theta \; \hat{\perp} - mg \sin \theta \; \hat{\parallel} \right) + mg \cos \theta \; \hat{\perp} + \gamma mg \cos \theta \; \hat{\parallel} \tag{5.10b}$$

$$= -mg (\sin \theta - \gamma \cos \theta) \; \hat{\parallel} \tag{5.10c}$$

According to equation 5.10c, the total force is then parallel to the surface. As we mentioned earlier, equation 5.10c would have a finite value when the pull force overcomes friction or if the term between parentheses is larger than zero.

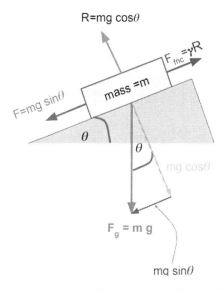

Figure 5.4: Free body diagram of an object placed on a tilted surface.

Wood on wood

Rubber on ice

Figure 5.5: Visualization of the minimum tilt angles needed to move a wooden object on a wooden surface and a rubber object on ice.

$$\overline{F}_{total} = \begin{cases} 0 & \sin\theta \leq \gamma\cos\theta \\ -mg(\sin\theta - \gamma\cos\theta)\,\hat{\parallel} & \sin\theta > \gamma\cos\theta \end{cases} \qquad (5.11)$$

The condition in equation 5.11 is obtained when the portion of the gravitational force that pulls the object downwards overcome friction, or

$$mg\sin\theta > \gamma mg\cos\theta \rightarrow \sin\theta > \gamma\cos\theta \qquad (5.12)$$

From equation 5.12, we can define the angle at which the object starts moving, θ_{min}, or the condition when gravitational pull just equates the friction.

$$\sin\theta_{min} = \gamma\cos\theta_{min} \rightarrow \gamma = \tan\theta_{min} \rightarrow \theta_{min} = \tan^{-1}\gamma \qquad (5.13)$$

Hence, the object starts moving once the surface is tilted at an angle larger than the inverse tangent of the friction coefficient between the object and surface materials. Table 5.1 shows some angles for different surfaces and objects.

Table 5.1: Examples of minimum title angle for different surfaces and objects.

Materials	Friction coefficient γ	Minimum angle θ_{min} ($^\circ$)
Rubber on asphalt (Dry)	0.68	34
Rubber on asphalt (wet)	0.4	21.8
Rubber on ice	0.15	8.5
Waxed ski on snow	0.05	2.9
Wood on wood	0.3	16.7
Steel on steel	0.57	29.7
Copper on steel	0.36	19.8

5C. Ramp pulley

Wheel lifting

Consider the ramp and wheel lifting device at figure 5.6. If we neglect friction, what would be the minimum weight that can hold the box from falling down the ramp? To be able to solve this problem we need to look at the free body diagram of the box. But before that we need to understand what the wheel, the rope and the attached weight do. The attached weight exerts a force that is equal to the gravitational force pointing downwards.

$$\overline{F}_{g,2} = -m_2 g \hat{y} \qquad\qquad (5.14)$$

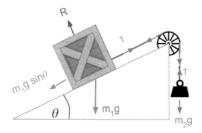

Figure 5.6: A free body diagram of a wheel lifting device.

This force causes a reaction in the rope which is commonly known as Tension $\overline{T}$, that points upwards towards the wheel. The box on the ramp as well is trying to pull the same rope down due to gravity and it is causing a reaction that equals the tension force in the rope. If we assume that the amplitude of the tension is T, we can write the following equations for the total force applied on the box

$$\overline{F}_1 = (T - m_1 g \sin\theta)\, \hat{\parallel} \qquad\qquad (5.15)$$

And the total force for the weight is

$$\overline{F}_2 = (T - m_2 g)\hat{y} \qquad\qquad (5.16)$$

We know that for any of the two objects, if the tension wins the gravitational force, the object will move upwards towards the wheel while the other object will be moving down away from the wheel. For the two objects to remain still, the total force on each should vanish, $|\overline{F}_1| = |\overline{F}_2| = 0$. This results in the following condition

$$T - m_1 g \sin \theta = T - m_2 g \rightarrow m_1 \sin \theta = m_2 \qquad (5.17)$$

From equation 5.17, we can say for the box to remain still on the ramp, the mass of the weight must equal $m_1 \sin\theta$.

Example 5.1:

If the ramp is tilted at 30º, the mass of the weight that is needed to keep the box from moving is?

$$m_2 = m_1 \sin(30^o) = \frac{1}{2} m_1$$

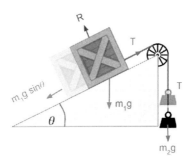

Figure 5.7: The gravitational force of the box wins the tension in the rope causing the box to move downwards.

Tension and acceleration

Let us now examine the situation when the total force is not necessarily zero in equations 5.15 and 5.16. In this case the two objects are accelerating. For the drawing in figure 5.7, the box is accelerating upwards, and the weight is accelerating downwards. It is not hard to guess that both objects are accelerating with the same amplitude. From the Newton's second law we know

$$\bar{a}_1 = \frac{\bar{F}_1}{m_1} = \left(\frac{T}{m_1} - g\sin\theta \right) \hat{\parallel} \qquad (5.18)$$

Similarly, for the weight we can obtain the following relation for its acceleration

$$\bar{a}_2 = -\left(g - \frac{T}{m_2}\right)\hat{y} \tag{5.19}$$

Notice that in equation 5.19 the negative sign is pulled outside to indicate that the weight is moving downwards. In the same scenario, the box is assumed to be accelerating up the ramp and hence the positive parallel direction is sustained. We have logically and based of observations predicted that the amplitude of the two accelerations should be equal, $|\bar{a}_1| = |\bar{a}_2| = a$. In this case

$$\frac{T}{m_1} - g\sin\theta = g - \frac{T}{m_2} \rightarrow T\left(\frac{1}{m_1} + \frac{1}{m_2}\right) = g(1 + \sin\theta) \tag{5.20}$$

Equation 5.20 can be rearranged to obtain the following relation for the tension

$$T = \frac{m_1 m_2 g(1+\sin\theta)}{m_1 + m2} \tag{5.21}$$

The acceleration can be written using equation 5.19 for instance as

$$a = g - \frac{T}{m_2} = g - \frac{m_1 g(1+\sin\theta)}{m_1 + m_2} \tag{5.22}$$

This can be simplified to

$$a = \frac{m_2 g - m_1 g \sin\theta}{m_1 + m_2} \tag{5.23}$$

- Tension: $T = \frac{m_1 m_2 g(1+\sin\theta)}{m_1 + m2}$
- Acceleration: $a = \frac{m_2 g - m_1 g \sin\theta}{m_1 + m_2}$

Adding friction

If the gravitational force wins the tension in the rope, what is the needed tilt angle for the box to move down the ramp when friction presents? This situation is presented in figure 5.8. The total force on the weight is the same as in equation 5.16. However, for the total force on the box we need to add the friction force. Equation 5.15 becomes

$$\overline{F}_1 = (T - m_1 g \sin\theta - \gamma m_1 g \cos\theta)\, \hat{\parallel} \tag{5.24}$$

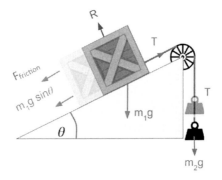

Figure 5.8: Adding friction to the wheel pulling a box that is placed on a tilted surface.

Using a similar approach as in the previous section by equating the acceleration amplitude

$$\frac{T}{m_1} - g\sin\theta - \gamma g\cos\theta = g - \frac{T}{m_2} \tag{5.25a}$$

$$\rightarrow T\left(\frac{1}{m_1} + \frac{1}{m_2}\right) = g(1 + \sin\theta + \gamma\cos\theta) \tag{5.25b}$$

The tension force is then

$$T = \frac{m_1 m_2 (1 + \sin\theta + \gamma\cos\theta)}{m_1 + m_2} \tag{5.26}$$

And the acceleration in this case is

$$a = \frac{m_2 g - m_1 g(\sin\theta + \gamma\cos\theta)}{m_1 + m_2} \tag{5.27}$$

What if the box's gravitational force wins the tension and the box moves down the ramp?

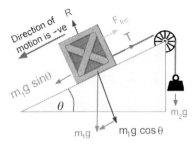

Figure 5.8: Motion of the box down the ramp with friction

The friction is in this case pointing at the positive parallel direction. Hence, the tension is

$$g \sin \theta - \gamma g \cos \theta - \frac{T}{m_1} = \frac{T}{m_2} - g \qquad (5.28a)$$

$$\rightarrow T = \frac{m_1 m_2 g (1 + \sin \theta - \gamma \cos \theta)}{m_1 + m_2} \qquad (5.28b)$$

The acceleration is

$$a = \frac{m_1 g (\sin \theta - \gamma \cos \theta) - m_2 g}{m_1 + m_2} \qquad (5.29)$$

Equations 5.27 and 5.29 show the acceleration when the weight wins the tension and pulls the box upwards and when the box moves downwards respectively. Let us examine the conditions when the box remains stationary without movement. In other words, when the acceleration equals zero. From equation 5.27.

$$a = 0 \rightarrow m_2 = m_1 (\sin \theta + \gamma \cos \theta) \qquad (5.30)$$

For the object to move up on the ram, the weight must have a mass

$$m_2 > m_1 (\sin \theta + \gamma \cos \theta) \qquad (5.31)$$

Using equation 5.29 we obtain the second limit

$$a = 0 \rightarrow m_2 = m_1 (sin\theta - \gamma cos\theta) \qquad (5.32)$$

For the box to move down the ramp, the box mass needs to satisfy the following

$$m_1(\sin\theta - \gamma\cos\theta) > m_2 \qquad\qquad (5.33)$$

Or

$$m_2 < m_1(\sin\theta - \gamma\cos\theta) \qquad\qquad (5.34)$$

Equations 5.31 and 5.34 show two conditions on the weight mass for which the object moves, either upwards or downwards. If the mass is however in between, then the box remains stationary, and the acceleration should equal zero. Based on this argument, the combined acceleration equation should be as follows

$$a =$$

$$\begin{cases} \frac{m_2 g - m_1 g(\sin\theta + \gamma\cos\theta)}{m_1 + m_2} & m_2 > m_1(\sin\theta + \gamma\cos\theta) \\ 0 & m_1(\sin\theta - \gamma\cos\theta) \leq m_2 \leq m_1(\sin\theta + \gamma\cos\theta) \\ \frac{m_1 g(\sin\theta - \gamma\cos\theta) - m_2 g}{m_1 + m_2} & m_2 < m_1(\sin\theta - \gamma\cos\theta) \end{cases}$$

$$(5.35)$$

The symbol T is used here for the tension force. We will however use the same symbol T to represent temperature and sometimes to represent the period as we will see letter in the oscillations chapter. The student shall not be confused as the meaning of the symbol will be easily identified through the context of the discussion.

5D. Summary

- For every action there is an equal and opposite reaction
  - A mass m on a flat surface at rest has
    - Gravitational force:
      - $\overline{F}_g = -mg\hat{y}$
    - Reaction force:
      - Normal to the surface
      - Opposite to gravitational force
      - $\overline{R} = mg\hat{y}$
    - Total force acting on the object:
      - $\overline{F}_g + \overline{R} = 0$
        - The object does not move
  - A mass m on a flat surface with force $F\hat{x}$ applied has:
    - Gravitational force: $\overline{F}_g = -mg\hat{y}$
    - Reaction force: $\overline{R} = mg\hat{y}$
    - Applied force: $\overline{F} = F\hat{x}$
    - Friction force:
      - Proportional to the amplitude of the reaction force
      - Opposite to the direction of the applied force
      - $\overline{F}_{friction} = -\gamma|\overline{R}|\hat{x}$
    - Total force acting on the object:
      - $\overline{F}_{total} = \overline{F}_g + \overline{R} + \overline{F} + \overline{F}_{friction} = (F - \gamma mg)\hat{x}$
      - The object moves when $F > \gamma mg$
  - A mass m on a tilted surface with angle θ and no force applied has:
    - Two directions of motion:
      - Parallel to the surface $\hat{\parallel}$
      - Normal to the surface $\hat{\perp}$
    - Gravitational force has two components:
      - Parallel: $-mg\sin\theta\ \hat{\parallel}$
      - Normal: $-mg\cos\theta\ \hat{\perp}$
    - Reaction force:
      - Normal to the surface
      - $\overline{R} = mg\cos\theta\ \hat{\perp}$

- Friction force:
  - Proportional to amplitude of reaction
  - Opposite to the parallel gravitational force component
  - $\overline{F}_{friction} = \gamma mg \cos\theta \; \hat{\mathbb{I}}$
- Total force acting on the object:
  - $\overline{F}_{total} = \overline{F}_g + \overline{R} + \overline{F}_{friction}$
    - Normal: $0 \; \hat{\perp}$
    - Parallel: $-mg(\sin\theta - \gamma\cos\theta) \; \hat{\mathbb{I}}$
  - The object moves when
    - Amplitude of parallel force > 0
    - $\sin\theta > \gamma\cos\theta$
  - Minimum tilt angle:
    - $\theta_{min} = \tan^{-1}\gamma$
- Free body diagram:
  - A diagram showing all the forces amplitude and directions that are applied on the object.
- Ramp pulley:
  - Connecting an object of mass m_1 on a tilted surface (ramp) with a rope
  - Connecting the other side of the rope with mass m_2
  - The rope goes through a wheel at the edge of the ramp
  - Forces on object 1:
    - Gravitational force:
      - Parallel: $-m_1 g \sin\theta \; \hat{\mathbb{I}}$
      - Normal: $-m_1 g \cos\theta \; \hat{\perp}$
    - Surface reaction:
      - $\overline{R} = mg \cos\theta \; \hat{\perp}$
    - No friction assumption
    - Tension in the rope:
      - $\overline{T} = T \; \hat{\mathbb{I}}$
    - Total force:
      - Parallel: $(T - \gamma mg \cos\theta) \; \hat{\mathbb{I}}$
  - Forces on object 2:
    - Gravitational force: $-m_2 g \hat{y}$
    - Tension in the rope: $\overline{T} = T \hat{y}$

- Total force: $(T - mg)\hat{y}$
- The objects remain steady when the total force on each is zero
  - Object 1:
    - $T = m_1 g \sin \theta$
  - Object 2:
    - $T = m_2 g$
  - $m_1 sin\theta = m_2$
- When objects move, they have accelerations:
  - Object 1:
    - $\overline{a}_1 = \left(\dfrac{T}{m_1} - g \sin \theta \right) \hat{\parallel}$
  - Object 2:
    - $\overline{a}_2 = \left(\dfrac{T}{m_2} - g \right) \hat{y}$
  - If Object1 moves down the ramp($-\hat{\parallel}$)
    - Object 2 moves up ($\hat{y}$)
  - If Object 1 moves up the ramp ($\hat{\parallel}$)
    - Object 2 moves down($-\hat{y}$)
  - $a1 = - a2$
    - $T = \dfrac{m_1 m_2 g (1 + \sin \theta)}{m_1 + m_2}$
- Adding friction:
  - If Object 1 moves upwards
    - Tension amplitude: $T = \dfrac{m_1 m_2 (1 + \sin \theta + \gamma \cos \theta)}{m_1 + m_2}$
    - Acceleration amplitude: $a = \dfrac{m_2 g - m_1 g (\sin \theta + \gamma \cos \theta)}{m_1 + m_2}$
    - Condition: $m_2 > m_1 (\sin \theta + \gamma \cos \theta)$
  - If Object 1 is moving downwards
    - Tension amplitude: $T = \dfrac{m_1 m_2 g (1 + \sin \theta - \gamma \cos \theta)}{m_1 + m_2}$
    - Acceleration amplitude: $a = \dfrac{m_1 g (\sin \theta - \gamma \cos \theta) - m_2 g}{m_1 + m_2}$
    - Condition: $m_2 < m_1 (\sin \theta - \gamma \cos \theta)$
  - The objects stay still when
    - $m_1 (\sin \theta - \gamma \cos \theta) \leq m_2 \leq m_1 (\sin \theta + \gamma \cos \theta)$

5E. Review questions

1.   Draw a free body diagram for an object placed on a tilted surface labeling all the force components.

2.   An object of mass of 500 grams and of unknown material is placed on a wooden table. When you titled the table by 14 degrees the object started to move. Could this object be possibly made of wood? Explain.

3.   In the lab, you were given an object and a surface then you were asked to calculate the friction coefficient between them without knowing their materials. Can you propose an experiment to measure the friction coefficient?

4.   Two copper boxes were placed on a steel surface. The boxes have masses of 2 kg and 3 kg respectively. The steel surface was tilted at $25°$. Which of these boxes will start moving? At what acceleration(s)?

5.   Friction is not considered in the wheel lift problem in figure 5.6, what are the needed weights to ensure that the box remains stationary?

Box mass m_1 (kg)	Tile angle ($°$)	Weight mass m_2 (kg)
3	20	
5	15	
1.5	30	

6.   Two boxes with masses 2 kg and 3 kg are placed at the two sides of the rope in figure 5.6, calculate the tension and the acceleration for the two possible combinations?

7.   A 2 kg steel box was placed on a 15 o titled steel surface. If a 1 kg weight was placed at the other end of the rope in figure 5.6, would the box move downward or upward or remain still?

8.   A wooden surface is tilted at 20 o. If a 4 kg wooden box is placed on the surface, what is the range of weights that can ensure that the box remains stationary?

9.   A copper sheet is placed on a table with a 2.5 kg steel object placed on it. The object is connected to a rope and the other end of the rope is connected to a 1 kg weight using a wheel at the edge of the table.

If we started tilting the table, at what angle would the object start to move?

CHAPTER SIX

CONSERVATION OF MOMENTUM AND ENERGY

6A. Conservation of momentum and collisions

Let us borrow the box from the previous chapter and place it on a flat frictionless surface. This way we set the friction force to zero. We then apply a force at one instant of time, t=0, then let the box move freely. The box then slides on the surface with a constant speed as no force is applied afterwards. We decided to place another box at the same surface in the path of the first box movement. What will happen when the moving box crashes into the still one?

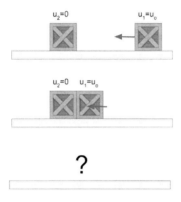

Figure 6.1: Collision between a moving box and one at rest.

It will exert a force on it. The second box reacts with an equal and opposite force according to Newton's third law of motion.

$$\overline{F}_2 = -\overline{F}_1 \tag{6.1}$$

We know from chapter 4, equation 4.2, that the momentum is the time change of the force.

$$\frac{d\overline{P}_1}{dt} = -\frac{\overline{P}_2}{dt} \tag{6.2}$$

Moving the right side in equation 6.2 to the left, we can write that

$$\frac{d\overline{P}_1}{dt} + \frac{d\overline{P}_2}{dt} = 0 \tag{6.3}$$

Then

$$\frac{d}{dt}(\overline{P}_1 + \overline{P}_2) = 0 \tag{6.4}$$

For the time derivative to equal zero, the term in parentheses in equation 6.4 should be constant.

$$\overline{P}_1 + \overline{P}_2 = constant \tag{6.5}$$

The total momentum is always constant. Hence, it should remain the same before the collision of the two boxes and after the collision. This is commonly referred to as the conservation of momentum.

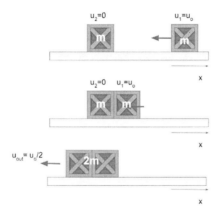

Figure 6.2: Two exact boxes, one sliding and the other is stationary.

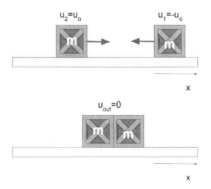

Figure 6.3: Two exact boxes sliding at opposite directions with the same speed.

Momentum and Collision

Newton's third law then indirectly states that the total momentum is always constant. Let us now assume that the two boxes in the example above each have a magnet attached to it. Because of these magnets, the two boxes will combine in one object with a mass that equals the sum of the two masses right after the collision. The conservation of momentum indicates that the total momentum before collision and after collision should be the same. Before collision, the first box with mass m_1 is moving in the positive x-direction with a constant speed u.

$$\overline{P}_1 = m_1 \cdot u\,\hat{x} \tag{6.6}$$

The second box is standing still before the collision, $\overline{P}_2 = 0$. The total momentum before collision is

$$\overline{P} = m_1 \cdot u\hat{x} + 0\hat{x} = m_1 \cdot u\hat{x} \tag{6.7}$$

After collision, the two boxes combine into one object of mass $m_1 + m_2$. The combined object should then move with a new speed u' along the positive x-direction.

$$\overline{P} = (m_1 + m_2)\overline{u}' \tag{6.8}$$

From the conservation of momentum, one knows that the momentum before collision and after collision should be equal.

$$(m_1 + m_2)u' = m_1 u \tag{6.9}$$

The speed of the combined objects after collision is

$$u' = \frac{m_1 u}{m_1 + m_2} \tag{6.10}$$

If the two boxes have the same mass $m_1 = m_2 = m$, then the speed after collision is

$$u' = \frac{1}{2}u \tag{6.11}$$

Hence, the combined object now moves with half the speed of the originally sliding box as illustrated in figure 6.2. If two exact boxes were originally sliding with the same speed but in opposite directions, then the total momentum is

$$\overline{P} = mu\hat{x} - mu\hat{x} = 0\hat{x} \tag{6.12}$$

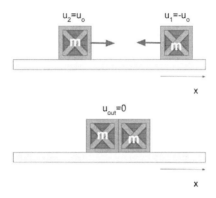

Figure 6.4: Collision between two objects in the same direction.

After collision, the combined object remains still as the momentum is zero as shown in figure 6.3. Now, what will be the output speed if the two objects were sliding at the same direction but with different speeds as in figure 6.4? In this case, the total momentum before collision is

$$\overline{P} = (m_1 \cdot u_1 + m_2 \cdot u_2)\hat{x} \tag{6.13}$$

The total momentum after collision

$$\overline{P} = (m_1 + m_2)u'\hat{x} \tag{6.14}$$

The output speed is then

$$u' = \frac{m_1 u_1 + m_2 u_2}{m_1 + m_2} \tag{6.15}$$

If the two objects have identical masses, then the output speed is $u' = \frac{1}{2}(u_1 + u_2)$ which is the average speed of the two objects. This type of collision is referred to as inelastic collision. This is because the mass of the objects changes during the collision, combined in one object. If the masses of the object do not change, then the collision is referred to commonly as elastic collision.

Collision without combining the objects

What if we did not place any special magnets or hooks to stick the boxes together, how can we predict their motion after collision? Before collision and after collision, the total momenta are:

$$\overline{P}_{before} = m_1 \cdot \overline{u}_1 + m_2 \cdot \overline{u}_2 \tag{6.16a}$$

$$\overline{P}_{after} = m_1 \cdot \overline{u}_1' + m_2 \cdot \overline{u}_2' \tag{6.16b}$$

Knowing that $\overline{P}_{before} = \overline{P}_{after}$, then

$$m_1 \cdot \overline{u}_1 + m_2 \cdot \overline{u}_2 = m_1 \cdot \overline{u}_1' + m_2 \cdot \overline{u}_2' \tag{6.17}$$

It is clear from equation 6.17 that to obtain the output velocities, the conservation of momentum alone is not sufficient to obtain a solution. We have two unknowns u_1' and u_2' and one equation. Hence, we need to investigate another quantity that is also conserved to get another equation. Luckily, we do have, and we do know that energy is also conserved.

6B. Conservation of energy

What was missing in the previous discussion is a very important law "conservation of energy." It is stated:

> "There is a certain quantity, which we call energy, that does not change in the manifold changes which nature undergoes."

What does that statement mean? It says that there is some numerical quantity (we call it energy) that does not change when something happens. (Statement from Feynman lectures on physics).

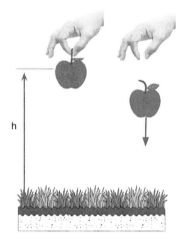

Figure 6.5: An apple falling to the ground under gravitational force.

Energy

Let's go back to the apple example. When we hold the apple at a height h from the ground it is at rest. The moment we open our hand, the apple starts to fall. This movement tells us that there is a sort of energy moving the apple down.

Now, we can think of this as follows:

When we hold the apple at height h, it has a form of stored energy (not causing any movement.) The moment we open our hand, the stored energy starts to transfer gradually into motion till the apple reaches the ground. We know from before that the gravitational force causes the apple to accelerate downwards once we leave it. Hence, the energy stored in the apple at height h is commonly referred to as the gravitational potential energy, U.

We can think of the selection of the word potential as when we hold the object at a certain height off the ground it has the potential to move downwards once we let it free. From observations we have noticed that holding a heavy object needs more work from us to lift it to the desired height compared to a lighter one. Hence, this energy depends on the weight, mg, and the height, h, of the object. Mathematically, we can say that

$$PE = mgh \qquad (6.17)$$

Work is the energy transferred to or from an object via the application of force along a certain displacement. Here, the force is the gravitational force, $|F| = mg$, and the displacement amplitude is h.

The unit of energy is Joule or J in short and it is $J = N \cdot m = kg \cdot m^2/s^2$. When the apple moves, it starts to gain a different form of energy that depends on how fast this object moves. Another sort of energy is now formed. This is referred to as the kinetic energy, KE.

From observations as well, we know that catching a moving tennis ball requires less work compared to catching a cannonball moving at the same speed. This is as the cannonball has a larger mass. It also requires more work to catch a fast tennis ball compared to a slow one. Hence, the kinetic energy depends on both the mass and the speed of the object.

$$KE = \frac{1}{2}mu^2 \qquad (6.18)$$

Conversation of energy

The law of conservation of energy indicates that the total energy shall remain constant all the time. Let us compare the energies in an apple in two situations: held at a height of h ($t = 0$) and during the free fall time (time = t).

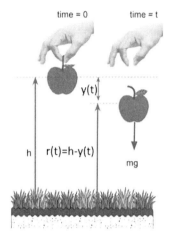

Figure 6.6: The apple under gravitational force at two different time instances.

Looking at table 6.1, one would ask a straightforward question, the total energies at the two-time instances do not seem to match. Would that violate the law of conservation of energy as we just stated? To answer this, we need to evaluate the displacement amplitude $r(t)$ and the speed $u(t)$ from Newton's second law of motion.

$$\overline{F} = -m \cdot \overline{a}(t), \text{ where } \overline{a}(t) = \frac{\overline{u}(t)}{dt} \tag{6.19}$$

From equation 6.19 we can write the following

$$\frac{d\overline{u}(t)}{dt} = -g\hat{y} \rightarrow \overline{u}(t) = -gt\hat{y} \tag{6.20}$$

The solution in equation 6.20 follows equation 2.4 in chapter two when setting the initial velocity to zero. Similarly, we can use equation 2.2 to evaluate the displacement vector knowing that $\overline{r}_o = h\,\hat{y}$ and the velocity is as in equation 6.20.

$$\overline{r}(t) = \left(h - \int_0^t gt\,dt\right)\hat{y} \tag{6.21a}$$

$$\overline{r}(t) = \left(h - \frac{1}{2}gt^2\right)\hat{y} \tag{6.21b}$$

Substituting the results from equations 6.20 and 6.21 into the total energy at time t, we obtain

$$PE(t) + KE(t) = mgr(t) + \frac{1}{2}mu^2(t) \tag{6.22a}$$

$$= mg\left(h - \frac{1}{2}gt^2\right) + \frac{1}{2}mg^2t^2 \tag{6.22a}$$

$$= mgh \tag{6.22c}$$

From the result in 6.22c, one can state that

$$PE(t) + KE(t) = mgh = PE(0) + KE(0) \tag{6.23}$$

Hence, the total energy remains the same at the two-time instances, which satisfies the law of conservation of energy.

Work

Let us pay a closer look at the hand's action when it throws a tennis ball up for serving as illustrated in figure 6.7. The hand applies a force on the ball over a displacement path/curve s (while the hand is grabbing the ball). This action causes the hand to transfer an amount of energy to the ball, so it moves. This amount of energy transferred to the ball is referred to as work.

"Work represents the change of the energy of the object due to the application of force along a displacement."

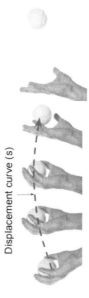

Figure 6.7: Energy transferred from the hand to the ball.

In the case of the apple held at a height h, when the hand leaves the apple gravity applies a constant force on it moving it by a distance $y(t)$. The work is then the amplitude of this force, mg, multiplied by the distance it was applied on, $y(t)$, or

$$W = mgy(t) \tag{6.24}$$

We can write the distance $y(t)$ as the difference between the amplitudes of the displacement vectors at time 0 and t, $y(t) = h - r(t)$.

$$W = mgy(t) = mgr(t) - mgrh \tag{6.25}$$

We know that the potential energies at $t = 0$ and t are $U(0) = mgh$ and $U(t) = mgr(t)$ respectively. Hence, we can rewrite equation 6.25 as

$$W = PE(t) - PE(0) \tag{6.26}$$

The work equals the change of the potential energy of the object. Notice that $y(t)$ in equation 6.24 is negative. That results in a negative value of the work. Negative sign indicates reduction of the potential energy. Hence, the produced work is taken from the potential energy. This work, that was taken from the potential energy, is produced as an increase in the kinetic energy

$$W = KE(t) - KE(0) = \frac{1}{2}mg^2t^2 = mg\left(\frac{1}{2}gt^2\right) \tag{6.27}$$

From equation 6.21b, we know that the amplitude of the displacement vector is

$$r(t) = h - \frac{1}{2}gt^2 \rightarrow \frac{1}{2}gt^2 = h - r(t) \tag{6.28}$$

Replacing equation 6.28 in equation 6.27

$$W = \Delta KE = mg(h - r(t)) = PE(0) - PE(t) = -\Delta PE \tag{6.29}$$

The work equals the increase in the kinetic energy (ΔKE) that equals the decrease in the potential energy ($-\Delta PE$).

Energy transfer

In the falling apple example, we observed that when the apple starts to fall its stored potential energy starts to decrease and its kinetic energy starts to increase. In other words, energy is transferred from the potential energy into kinetic energy. Figure 6.8 below visualizes the energy transfer in archery, when the hand pulls the string and when the arrow is released.

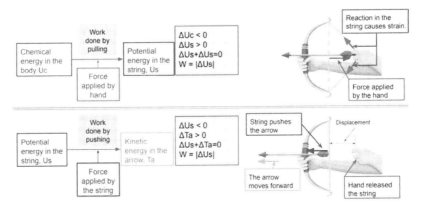

Figure 6.8: Visualization of the energy transfer between the hand, the string and the arrow in archery.

- The total energy is conserved, $PE + KE = constant$.
- The change of the total energy is zero, $\Delta(PE + KE) = 0$.
- Then, $\Delta PE + \Delta KE = 0 \rightarrow \Delta KE = -\Delta PE$. The increase in kinetic energy equals the decrease in the potential energy.
- The work equals the change in potential energy, $W = \Delta PE$

6C. Torque and rotation

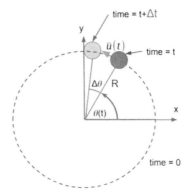

Figure 6.9: Circular motion of a ball around the center of motion

Work and rotation

What about rotation? How much work is needed to rotate the ball attached to a rope with an angle of $\Delta\theta$. We have learned that work is the change of energy. For the ball rotating in figure 6.10, the kinetic energy at any point of time is

$$KE = \frac{1}{2}m|\overline{u}|^2 = \frac{1}{2}m\left(u_x^2 + u_y^2\right) \qquad (6.30)$$

The change of the kinetic energy is

$$\Delta KE = \frac{1}{2}m\left(2u_x\Delta u_x + 2u_y\Delta u_y\right) \qquad (6.31)$$

We know that the change of speed per time gives the acceleration. Hence,

$$\frac{\Delta u_x}{\Delta t} = a_x \rightarrow \Delta u_x = a_x \cdot \Delta t \qquad (6.32a)$$

$$\frac{\Delta u_y}{\Delta t} = a_y \rightarrow \Delta u_y = a_y \cdot \Delta t \qquad (6.32b)$$

Substituting into equation 6.31

$$\Delta KE = ma_x(u_x\Delta t) + ma_y\left(u_y\Delta t\right) \qquad (6.33)$$

We know from Newton's second law of motion that force equals mass multiplied by acceleration. Then equation 6.33 becomes

$$\Delta KE = F_x \cdot u_x \Delta t + F_y \cdot u_y \Delta t \qquad (6.34)$$

Using equations 2.27 in chapter two, we can rewrite equation 6.34 in terms of the angular frequency as

$$\Delta KE = F_x(-R\Omega \cdot \Delta t \cdot \sin\theta) + F_y(R\Omega \cdot \Delta t \cdot \cos\theta) \qquad (6.35)$$

We know that $\theta = \Omega t \rightarrow \Delta\theta = \Omega \cdot \Delta t$. Also, from equations 2.24 in chapter two we are aware that $R \cdot \sin\theta = y$ and $R \cdot \cos\theta = x$. Hence, equation 6.35 can be rewritten as

$$\Delta KE = (-y \cdot F_x + x \cdot F_y)\Delta\theta \qquad (6.36)$$

Rise of torque

Equation 6.36 give us the amount of work needed to turn an object by an angle $\Delta\theta$ around a center of rotation. The ratio of the work done to the angle of rotation is defined as the torque.

Torque means twist, which is a twisting force that tends to cause rotation.

based on equation 6.36, we can derive an expression for torque that is work per unit angle, $\tau = \frac{\Delta KE}{\Delta\theta}$ as

$$\tau = -y \cdot F_x + x \cdot F_y \qquad (6.37)$$

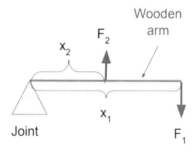

Figure 6.10: Two forces applied on a wooden arm mounted to a joint.

For the example in figure 6.10, There exists more than one force applied onto the rod. What is the total torque?

The first torque is: $\tau_1 = -x_1 \cdot F_1$

The first torque is: $\tau_2 = x_2 \cdot F_2$

The total torque is:

$$\tau = \tau_1 + \tau_2 = -x_1 \cdot F_1 + x_2 \cdot F_2 \tag{6.38}$$

The negative sign is since F_1 is pointing down. For the arm in the figure not to move, the total torque should equal zero. That means that there is no change in the stored energy in the system.

$$\tau = 0 = -x_1 \cdot F_1 + x_2 \cdot F_2 \rightarrow \frac{F_1}{F_2} = \frac{x_2}{x_1} \tag{6.39}$$

Torque and balance

For the balance in figure 6.12 with two masses: m_1 and m_2, the scale will be balanced horizontally when the total torque is zero. Knowing that both forces applied on the balance are gravitational pull, then both forces are pointing downwards.

$$\tau = 0 = -x_1 \cdot m_1 g - x_2 \cdot m_2 g \tag{6.40}$$

If the origin of the x axis is set to the middle of the balance, then x_1 will have a negative value, $x_1 = -|x_1|$ and x_2 is positive, $x_2 = |x_2|$. Then the condition in equation 6.40 becomes

$$\frac{m_1}{m_2} = \frac{|x_2|}{|x_1|} \tag{6.41}$$

For the balance system at the bottom of figure 6.12, the needed ratio between the masses to keep the balance is $\frac{m_2}{m_1} = \frac{2L}{L}$ and hence $m_2 = 2m_1$.

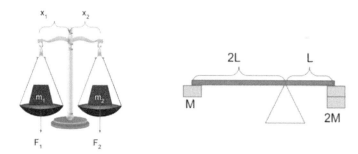

Figure 6.11: Balance of different weights and arm length. On the right, the length of the smaller mass needs to be larger to keep the balance.

6D. Conservation of energy and collision

Inelastic collision

Let us now revisit the collision case where the two objects combine and examine the kinetic energy before and after the collision. From equation 6.15, we know that if two objects of speeds u_1 and u_2 and masses m_1 and m_2 collide then the combined object has a mass of $m_1 + m_2$ and speed $u' = \frac{m_1 u_1 + m_2 u_2}{m_1 + m_2}$. The total kinetic energy before collision is

$$KE_{before} = \frac{1}{2} m_1 u_1^2 + \frac{1}{2} m_2 u_2^2 \qquad (6.42)$$

The total kinetic energy after the collision is

$$KE_{after} = \frac{1}{2}(m_1 + m_2) \left(\frac{m_1 u_1 + m_2 u_2}{m_1 + m_2} \right)^2 \qquad (6.43)$$

As we can see from equations 6.42 and 6.43 the energy before and after collision is not the same. If the second object was standing still before collision, $u_2 = 0$, then the energies before and after are $KE_{before} = \frac{1}{2} m_1 u_1^2$ and $KE_{after} = \frac{1}{2} \frac{m_1^2 u_1^2}{m_1 + m_2}$. The energy after the collision is less than before. So, where did the rest of the energy go? The answer is that the rest of the energy is consumed in combining the two objects together. Hence, the collision is considered inelastic as part of the energy is lost during the collision.

Elastic collision

Now, consider that the collision does not change the objects as in figure 6.13. In other words, the two objects do not combine or lose part of their mass. In this case we apply the conservation of energy between the kinetic energies before and after the collision, $KE_{before} = KE_{after}$.

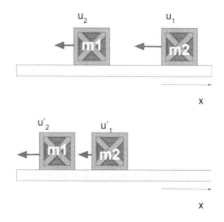

Figure 6.12: Elastic collision of two objects.

Conservation of momentum:

$$m_1 u_1 + m_2 u_2 = m_1 u_1' + m_2 u_2' \tag{6.44}$$

Conservation of energy:

$$m_1 u_1^2 + m_2 u_2^2 = m_1 {u_1'}^2 + m_2 {u_2'}^2 \tag{6.45}$$

We can rearrange equation 6.45 as follows

$$m_1 (u_1^2 - {u_1'}^2) = m_2 ({u_2'}^2 - u_2^2) \tag{6.46}$$

Expanding the term in parentheses knowing that

$$a^2 - b^2 = (a - b)(a + b) \tag{6.47}$$

$$m_1 (u_1 - u_1')(u_1 + u_1') = m_2 (u_2' - u2)(u_2' + u_2) \tag{6.48}$$

We can rearrange the conservation of momentum in equation 6.44 to be

$$m_1 (u_1 - u_1') = m_2 (u_2' - u_2) \tag{6.49}$$

Using this relation, equation 6.48 is simplified to

$$u_1 + u_1' = u_2 + u_2' \tag{6.50}$$

Solving equations 6.50 and 6.49 we obtain

$$u_1' = \frac{m_1 - m_2}{m_1 + m_2} u_1 + \frac{2m_2}{m_1 + m_2} u_2 \qquad (6.51\text{a})$$

$$u_2' = \frac{2m_1}{m_1 + m_2} u_1 + \frac{m_2 - m_1}{m_1 + m_2} u_2 \qquad (6.51\text{b})$$

Example 6.1:

Consider the case when the two boxes have the same mass, and the second box was originally at rest.

$m_1 = m_2 = m$ and $u_2 = 0$

$$u_1' = \frac{m - m}{m + m} \cdot u_1 + \frac{2m}{m + m} \cdot 0 = 0 \qquad (6.52\text{a})$$

$$u_2' = \frac{2m}{m + m} \cdot u_1 + \frac{m - m}{m + m} \cdot 0 = u_1 \qquad (6.53\text{b})$$

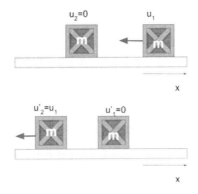

Figure 6.13: Elastic collision between a moving object and a steady one.

The collision causes the first box to stop and the second box to move with the original speed as the first box. Hence, the kinetic energy remains the same before and after the collision.

Example 6.2:

Both boxes have the same mass and the same speed but in opposite directions

$m_1 = m_2 = m$ and $u_2 = -u_1$

$$u_1' = \frac{m-m}{m+m} \cdot u_1 - \frac{2m}{m+m} \cdot u_1 = -u_1 \qquad (6.54a)$$

$$u_2' = \frac{2m}{m+m} \cdot u_1 - \frac{m-m}{m+m} \cdot u_1 = u_1 \qquad (6.54b)$$

The collision causes the two boxes to switch directions. They bounce back with the same speed.

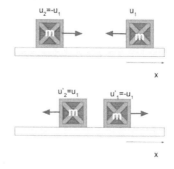

Figure 6.14: Elastic collision between two identical objects moving in opposite directions.

6E. Summary

- Conservation of momentum:
    - Momentum never changes on isolated collision of the objects
    - Collision of two objects:
        - $\overline{P}_1 + \overline{P}_2 = constant$
            - Momentum of object 1: $\overline{P}_1 = m_1 \cdot \overline{u}_1$
            - Momentum of object 2: $\overline{P}_2 = m_2 \cdot \overline{u}_2$
            - Total momentum: $\overline{P} = m_1 \cdot \overline{u}_1 + m_2 \cdot \overline{u}_2$
    - Inelastic collision:
        - The two objects combine after collision
        - Momentum of combined object: $\overline{P} = (m_1 + m_2)u'$
        - The velocity after collision: $\overline{u}' = \left(\frac{m_1}{m_1+m_2}\right)\overline{u}_1 + \left(\frac{m_2}{m_1+m_2}\right)\overline{u}_2$
        - If $m_1 = m_2$, then: $\overline{u}' = \frac{1}{2}(\overline{u}_1 + \overline{u}_2)$
- Conservation of energy:
    - Energy does not change when something happens.
    - Two types of energy:
        - Potential energy PE
            - Energy stored in the system
        - Kinetic energy KE
            - Energy due to motion
        - Energy of the system is always constant
            - $PE + KE = constant$
    - In case of gravity
        - An object of mass m that is held at height h has
            - Potential energy:
                - $PE = mgh$
                - That is the work needed to lift the object to height h.
            - Kinetic energy:
                - $KE = 0$
                - The object is not moving
            - Total energy:
                - $PE + KE = mgh$

- The object is left to fall
  - Acceleration: $\bar{a}(t) = -g\hat{y}$
  - Velocity: $\bar{u}(t) = -gt\hat{y}$
  - Displacement: $\bar{r}(t) = \left(h - \frac{1}{2}gt^2\right)\hat{y}$
  - Potential energy: $PE = mgr(t)$
    - $PE = mgh - \frac{1}{2}mg^2t^2$
  - Kinetic energy: $KE = \frac{1}{2}mu(t)^2$
    - $KE = \frac{1}{2}mg^2t^2$
  - Total energy:
  - $PE + KE = mgh$
- Work:
  - The energy of the object due to the application of force along a displacement.
    - In case of gravity
      - Change of potential energy from h to height r(t)
        - $W = mg(r(t) - h)$
        - $W = PE(t) - PE(0) = \Delta PE$
        - $W = -\frac{1}{2}mg^2t^2$
      - Change of kinetic energy from zero speed to u(t)
        - $\Delta KE = KE(t) - KE(0)$
        - $\Delta KE = \frac{1}{2}m \cdot u(t)^2 - 0$
        - $\Delta KE = \frac{1}{2}mg^2t^2 = -\Delta PE$
      - The produced work by reducing the potential energy increases the kinetic energy of the object while falling.
      - Energy transfers from one form to another but the total energy remains constant.
- Torque:
  - It means twist
  - Measure of force needed to make an object to rotate around an axis.
  - The rate of change of kinetic energy per rotational angle
    - $\tau = \frac{\Delta KE}{\Delta \theta} = -y \cdot F_x + x \cdot F_y$
  - When forming a balance
    - Object 1 has:
      - Mass: m_1 and location: $x_1 = |x_1|$

- Object 2 has:
  - Mass: m_2 and location: $x_2 = -|x_2|$
- Balance means $\tau = 0$
  - $0 = -m_1 g \cdot |x_1| - m_2 g \cdot (-|x_2|)$
  - $\frac{m_1}{m_2} = \frac{|x_2|}{|x_1|}$
- Conservation of energy and collision inelastic collision
  - Kinetic energy:
    - Before collision:
      - $KE_{before} = \frac{1}{2} m_1 u_1^2 + \frac{1}{2} m_2 u_2^2$
    - After collision:
      - $KE_{after} = \frac{1}{2}(m_1 + m_2) u'^2$
    - The difference in the energy goes in combining the objects together
- Elastic collision
  - Objects are moving along the x-axis
    - Conservation of momentum
      - $m_1 u_1 + m_2 u_2 = m_1 u_1' + m_2 u_2'$
    - Conservation of energy
      - $m_1 u_1^2 + m_2 u_2^2 = m_1 u_1'^2 + m_2 u_2'^2 \rightarrow u_1 + u_1' = u_2 + u_2'$

6F. Review questions

1. Two identical boxes placed on a frictionless surface. One box was given a force such that it slides towards the stationary box with a speed of 2 m/s. What is the speed of the combined object resulting from the collision?

2. Two boxes of mass 2 kg and 3 kg were placed on a frictionless surface. The two boxes are given a force such that they slide towards each other. If the heavier box speed is 3 m/sec, what is the speed needed for the second box such that the combined object after collision has a total speed of zero.

3. Complete the following table for inelastic collision between two objects which combine after the collision.

Box 1 mass (kg)	Box 2 mass (kg)	Box 1 velocity (m/s)	Box 2 velocity (m/s)	Combined object velocity (m/s)
2	0.5	$2x$	$1x$	
1	1	$4x$		$0x$
3		$-2x$	$2x$	$-1x$
	5	$0x$	$2x$	$1.5x$
3	1		$-1.5x$	$1x$

4. Calculate the work performed on a ball of mass 1 kg by gravity to free fall from a height of 5 m to a height of 2 m? If the ball was held steady before falling, how long would it take for the ballet to reach the 2m height and what would be its speed? What is the kinetic and potential energies of the ball at both heights?

h

Figure 6.15: A bodybuilder lifting weight.

5.   A bodybuilder is lifting a weight to a height h as in figure 6.15.

   1.   What is the work needed by the bodybuilder to lift 20 kg to a height of 30 cm?
   2.   If the weight increases to 40 cm, what would be the height the person lifts if he does the exact same amount of work as in (a).
   3.   If the rope at the bodybuilder side is tilted at an angle of 15 o, draw the free body diagram of the system and what is the minimum force needed to keep a 50 kg mass steady?

6.   A 500 gram toy car is moving at a constant speed of 2 m/s. What is the work needed by your hand in order to stop the car?

7.   For the hand lifting the tennis ball in figure 6.7, if the tennis ball was lifted to a height of 80 cm and it was given a constant speed of 4 m/s, what is the total work the hand had to perform?

8.   For the system in figure 6.10, complete the table below such that the wooden stick does not rotate.

x_1(m)	x_2(m)	F_1 (N)	F_2 (N)
0.4	0.8		2
	2	3	6
1		4	10
0.5	1.2	1.5	

9.   For the balance in the bottom in figure 6.11, if the arm length of the
      left side increases from 2L to 3L and the mass at the right side
      remains the same (2M), what would be the needed mass in the left
      side to keep the balance?

10.   For the elastic collision in figure 6.12, complete the table below

m_1(kg)	m_2(kg)	u_1(m/s)	u_2(m/s)	u_1'(m/s)	u_2'(m/s)
2	2	4	-4		
3	4	2	-2		
5		2	-3	0	
	2		0	2	1

CHAPTER SEVEN

OSCILLATIONS

7A. Spring and oscillations

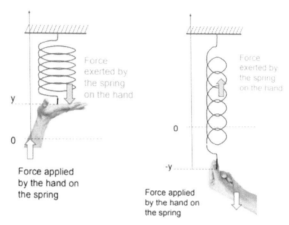

Figure 7.1: Force felt by the spring when it is being pushed or pulled.

In chapter three, we studied Hooke's law that describes the force on a spring. We have learned that when one pulls the spring in figure 7.1 by a distance y, then an opposite force is felt trying to push the spring back with the amplitude

$$\overline{F} = -K \cdot y\, \hat{y} \tag{7.1}$$

The constant K in equation 7.1 is the spring constant and it has units of N/m. What we did not study back then is what would happen if we let the spring go. For example, we start from the situation at the right side in figure 7.1, then at one instant of time, marked as t=0, we open our hand letting the spring free and keep track of the location of the spring end versus time. We know based on our experience and general knowledge that the spring will move up passing the y = 0 point then stops and bounces back in a repeated fashion. This periodical motion of the spring is commonly known as oscillation.

Simple harmonic oscillation

Let us put some mathematical representations for the spring motion. To know what will happen in time, we need to use Newton's second law of motion. Assume that we place an object of mass m at the tip of the spring and let us neglect the mass of the spring. We could as well neglect the effect of gravity and friction caused by air resistance or the connection between the spring and well. In this case we can write Newton's second law as follows

$$\overline{F} = m\overline{a} \tag{7.2}$$

With the assumptions we made, the force on the mass comes only from Hooke's laws as presented in equation 7.1.

$$-K \cdot y \, \hat{y} = m\overline{a} \rightarrow \overline{a} = -\frac{K}{m} y \, \hat{y} \tag{7.3}$$

The acceleration as was introduced in chapter two, is the second derivative of time of the displacement vector, $\overline{a} = \frac{d^2\overline{r}}{dt^2}$. Since the motion is expected to be along the y access then we can consider only the amplitude of the acceleration vector, a, keeping in mind that its direction is along the y axis.

$$a(t) = \frac{d^2y}{dt^2} \tag{7.4a}$$

Hence,

$$\frac{d^2y}{dt^2} = -\frac{K}{m} y \tag{7.4b}$$

So, what is the solution to equation 7.4b. The equation states that the second derivative of the location y in time equates the negative of a constant multiplied by itself. One solution of this second order ordinary differential equation is in the form

$$y(t) = A \cdot \cos(\omega_o \cdot t) + B \cdot \sin(\omega_o \cdot t) \tag{7.5}$$

Hare, A, B and ω_o are constants. The second derivative of equation 7.5 with respect to time is obtained as follows

$$\frac{dy(t)}{dt} = -\omega_o \cdot A \cdot \sin(\omega_o \cdot t) + \omega_o \cdot B \cdot \cos(\omega_o \cdot t) \tag{7.6a}$$

$$\frac{d^2y(t)}{dt^2} = \frac{d}{dt}\left(\frac{dy(t)}{dt}\right) = -\omega_o^2 \cdot A \cdot \cos(\omega_o \cdot t) - \omega_o^2 \cdot B \cdot \sin(\omega_o \cdot t) \tag{7.6b}$$

Using the original representation of y in equation 7.5, we can write that

$$\frac{d^2y(t)}{dt^2} = -\omega_o^2 \cdot y(t) \tag{7.7}$$

Equation 7.7 above is the same as equation 7.4b when $\omega_o = \sqrt{\frac{K}{m}}$. Hence, we can write the following solution for the location of the mass versus time.

$$y(t) = A \cdot \cos\left(\sqrt{\frac{K}{m}} \cdot t\right) + B \cdot \sin\left(\sqrt{\frac{K}{m}} \cdot t\right) \qquad (7.8)$$

What about the constants A and B? We know from the statement of the problem that initially (at time $t = 0$) the spring was at a location $-y_o$ as illustrated in figure 7.1. So, if we set the time to zero ($t = 0$) and $y(0) = -y_o$ in equation 7.8 we obtain the following

$$y(0) = -y_o = A \cdot \cos(0) + B \cdot \sin(0) \qquad (7.9)$$

We all know by now that $cos(0) = 1$ and $sin(0) = 0$. That gives a solution $A = -y_o$. If we now, consider the situation when a time $t = \frac{\pi}{\sqrt{K/m}}$ has elapsed. In this case the term inside the cosine and sine will equal π. From the math review in chapter one, we should know that $\cos(\pi) = -1$ and $\sin(\pi) = 0$. Hence, equation 1.9 becomes

$$y(t_\pi) = -y_o \cdot \cos(\pi) + B \cdot \sin(\pi) = y_o \qquad (7.10)$$

This is the maximum distance that the spring could bounce to. Here $t_\pi = \frac{\pi}{\sqrt{K/m}}$ is the time it takes to move from location $-y_o$ to y_o. Now, what about the second constant, B?

Front the illustration in figure 7.1, we could notice that initially the hand was holding the mass. Hence, the mass was steady with a zero initial speed, the time derivative of distance. So, we only need to place $t = 0$ in equation 7.6a while setting the derivative to zero

$$\frac{dy(0)}{dt} = 0 = \omega_o \cdot B \qquad (7.11)$$

That gives us a solution, B = 0. Hence, the location of the mass m in time when it was pulled from the equilibrium position, set at the origin, to an initial location $-y_o$ is expressed as

$$y(t) = -y_o \cdot \cos\left(\sqrt{\frac{K}{m}} \cdot t\right) \qquad (7.12)$$

This motion is depicted in figure 7.2. One can notice from the graph that after a time of $2t_\pi$ the mass returns to the initial location or $y = -y_o$. This time is commonly referred to as one complete cycle or a time period T.

$$T = 2t_\pi = \frac{2\pi}{\sqrt{K/m}} \tag{7.13}$$

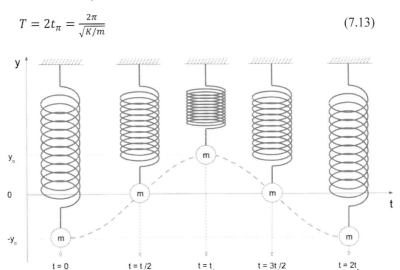

Figure 7.2: Oscillation of a mass attached to a sprint in one complete cycle.

After completing a cycle, the spring will repeat the same motion again where, by the end of each cycle the mass returns back to the initial location. The location versus time is graphically illustrated in figure 7.3. This type of motion is commonly referred to as simple harmonic oscillation.

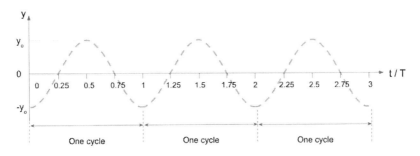

Figure 7.3: The location of the mass during three complete cycles.

Another approach of derivation

We could have used a different proposed solution of the differential equation 1.4b to reach the same solution. If we select
$$y(t) = A \cdot e^{\alpha t}$$

Then substituting this solution into equation 7.4b to obtain the following relation
$$\alpha^2 \cdot A \cdot e^{\alpha t} = -A \cdot \frac{K}{m} \cdot e^{\alpha t}$$
Dividing both sides by $A \cdot e^{\alpha t}$ we reach
$$\alpha^2 = -\frac{K}{m} \rightarrow \alpha = \pm i \sqrt{\frac{K}{m}}.$$
As we obtain two solutions $\alpha = i\sqrt{\frac{K}{m}}$ and $\alpha = -i\sqrt{\frac{K}{m}}$, then we write the displacement as summation of the two solutions. If we define $\omega_o = \sqrt{\frac{K}{m}}$ then the location versus time is
$$y(t) = A \cdot e^{i\omega_o t} + B \cdot e^{-i\omega_o t}$$

Here, A and B are generally different constants. Now, we apply the initial conditions. First, we know that the initial speed is zero, $u(0) = \frac{dy(0)}{dt} = 0$, hence
$$i\omega_o \cdot A - i\omega_o \cdot B = 0 \rightarrow A = B$$
And the object displacement equation becomes
$$y(t) = A \cdot (e^{i\omega_o t} + e^{-i\omega_o t})$$
From the mathematical review chapter one, we can easily identify that the term in brackets equals $2\cos(\omega_o t)$. Finally, we know that at time t=0, the spring was at location, $y(0) =- y_o$. That gives $A = -y_o/2$. Hence, the displacement become
$$y(t) = -y_o \cdot \cos(\omega_o t) = -y_o \cdot \cos(\sqrt{K/m} \cdot t) - y_o = 2A, \text{ hence}$$
$$y = -y_o \cdot \cos(K/mt)$$

and that is the result we obtained earlier.

Period, phase and frequency

The plot in figure 7.3 shows the oscillation of a mass m attached to the spring due to an initial pull to location $-y_o$ while neglecting the gravitational and friction effects. In each period $T = \frac{2\pi}{\omega_o} = \frac{2\pi}{\sqrt{K/m}}$ the spring completes one cycle and returns to its initial location $y = -y_o$. If a time t has elapsed, how could we tell the number of cycles that the mass has completed? To answer this logically, we use the information in our hands. We know that each period T, one complete cycle elapses. Hence, at a time t, there are $N = t/T$ cycles that have been completed. So, for 1 second, we say there are $f = 1/T$ completed cycles. We generally define the number of cycles per one second as the frequency of the oscillation.

$$f = \frac{1}{T} = \frac{\sqrt{K/m}}{2\pi} \tag{7.14}$$

The frequency has units of 1/sec or Hertz, Hz in short. Sometimes it is common to use the angular frequency $\omega_o = \frac{2\pi}{T} = 2\pi f$ instead of the frequency. ω_o has units of radian per second. Using those notations, we can rewrite equation 7.12 as

$$y(t) = -y_o \cdot cos(\omega_o t) = -y_o \cdot cos(2\pi f t) = -y_o \cdot cos\left(2\pi \frac{t}{T}\right) \tag{7.15}$$

So, every time a cycle is completed, the term inside the cosine, $2\pi \frac{t}{T}$, increases by a value of 2π. We can write that mathematically as

$$y(t + T) = -y_o \cdot cos\left(2\pi \frac{t+T}{T}\right) = -y_o \cdot cos\left(2\pi \frac{t}{T} + 2\pi\right) \tag{7.16a}$$

$$= -y_o \cdot cos\left(2\pi \frac{t}{T}\right) = y(t) \tag{7.16b}$$

Hence, the position of the spring returns to its initial place after one cycle or multiple complete cycles. We write that mathematically as

$$y(t + mT) = y(t), m = 0,1,2,3,... \tag{7.17}$$

The term inside the cosine is commonly called the phase, $\phi(t) = 2\pi f t = 2\pi \frac{t}{T}$. So, every complete cycle the phase changes by 2π, or

$$\phi(t + mT) = 2\pi \frac{t+mT}{T} = \phi(t) + 2\pi m = 0,1,2,3,.... \tag{7.18}$$

Phase has units of radians.

Natural frequency

Notice that in the equation of the spring oscillation in 7.12, the frequency at which the attached mass oscillates does not depend on the initial displacement. Initial displacement only affects the strength of the oscillation. Instead, the frequency ω_o only depends on the spring constant K and the mass m. Figure 7.4 shows the oscillation of the spring when pulled to different initial displacements.

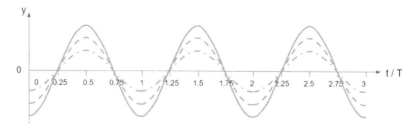

Figure 7.4: Oscillation of the mass m attached to the spring with spring constant K at different initial displacements.

The figure clearly shows that no matter how hard you pull the spring, it will oscillate at the same frequency, $\omega_o = \sqrt{\dfrac{K}{m}}$. This is commonly known as the natural frequency of the system. It is the frequency at which the system tends to oscillate even when no external force is present. The object motion under this simple harmonic oscillation can be summarized in table 7.1. The table shows the relations for three scalar quantities: displacement, speed and acceleration. Scalar quantities were considered as the direction of oscillation is assumed to be along the y axis as stated earlier.

Table 7.1: Summery of the simple harmonic oscillation motion.

Quantity	Relation	Equation	Units
Displacement	$y(t)$	$y(t) - y_o \cdot cos(\omega_o t)$	m
Speed	$u(t) = \dfrac{dy(t)}{dt}$	$u(t)$ $= \omega_o \cdot y_o \cdot sin(\omega_o t)$	m/s
Acceleration	$a(t) = \dfrac{d^2 y(t)}{dt^2}$	$a(t)$ $= \omega_o^2 \cdot y_o \cdot cos(\omega_o t)$	m/s^2

The energy of the system

As explained earlier, there are two governing energies in any system we deal with namely potential energy and kinetic energy. Potential energy is the energy stored in the object that once released it transfers partially or totally into kinetic energy. Gravity is an example. In our oscillating spring, it is the energy that the spring has while being pressed or stretched. We can write this mathematically as

$$PE = \int_0^y |\overline{F}| dy = \int_0^y Ky dy = \frac{1}{2} K y^2 \qquad (7.19)$$

We know that the position y changes with time as $y(t) = -y_o \cdot cos(\omega_o t)$, hence

$$PE = \frac{1}{2} K y_o^2 cos^2(\omega_o t) \qquad (7.20)$$

The kinetic energy on the other hand depends on the speed of the object at any moment of time,

$$KE = \frac{1}{2} m u(t)^2 \qquad (7.21)$$

Using the speed of the mass versus time in table 7.1, then the kinetic energy becomes

$$KE = \frac{1}{2} m y_o^2 \omega_o^2 \cdot sin^2(\omega_o t) \qquad (7.22)$$

So, the total energy of the system is the summation of the potential and kinetic energies

$$E = PE + KE = \frac{1}{2} K y_o^2 \cdot cos^2(\omega_o t) + \frac{1}{2} m \omega_o^2 y_o^2 \cdot sin^2(\omega_o t) \quad (7.23a)$$

$$E = \frac{1}{2} y_o^2 K \left(cos^2(\omega_o t) + \frac{m \omega_o^2}{K} sin^2(\omega_o t) \right) \qquad (7.23b)$$

Front the definition of the natural frequency, $\omega_o = \sqrt{K/m} \rightarrow \omega_o^2 = K/m$. That makes the ratio $m\omega_o^2/K = 1$. Hence, equation 7.23b becomes

$$E = \frac{1}{2}Ky_o^2(\cos^2(\omega_o t) + \sin^2(\omega_o t)) \qquad (7.24)$$

We already know from chapter one that $\cos^2\theta + \sin^2\theta = 1$. Then the total energy expression is reduced to

$$E = \frac{1}{2}Ky_o^2 \qquad (7.25)$$

The total energy of the system is then constant and depends on the initial displacement that we introduced in the spring. That result is expected due to the conservation of energy law we discussed in the previous chapter. Figure 7.4 visualizes the relationship between kinetic and potential energies of the spring over one complete cycle. As we can see in the figure at time t=0, the spring was at a maximum stretch, but it was not moving, hence the total energy is pure potential energy. When the hand leaves the spring, the mass m starts to gain speed and hence, its kinetic energy increases till the mass reaches the point of origin where there is no displacement, $y = 0$. That is the point where the string does not experience any stretching or pressing. At that point the total energy in the system is purely kinetic energy. Once the mass passes this point upwards, the spring starts to be pressed and hence an opposite force starts to work against the movement of the mass. As a result, the kinetic energy starts to reduce, and the potential energy starts to increase. This continues until the mass reaches a maximum displacement, $y = y_o$. At that point the mass stops and the kinetic energy vanishes. The mass starts moving downwards after that and reaches another stop at which the cycle is completed as shown in the figure. It is worth remembering always that at any moment during the oscillation the total energy which is the sum of both energies remain constant.

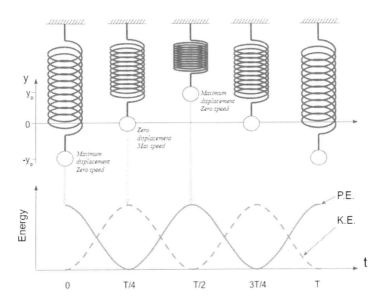

Figure 7.4: The change of potential and kinetic energies over one complete cycle of oscillation.

7B. Effect of gravity and damping force

Let's see what happens when we include gravity. For that, we would need to revisit the force equation in 7.3 and add the gravitational force to it.

Effect of gravity

We know that gravity always pulls the mass downwards in the -y direction. Hence, we can write the following

$$-Ky\hat{y} - mg\hat{y} = m\bar{a} \tag{7.26}$$

Dividing by the mass and assume the acceleration is along $\hat{y}$ then

$$\frac{d^2y(t)}{dt^2} = -g - \frac{K}{m}y \tag{7.27}$$

To solve equation 7.27, we could use a simple trick. We can arrange the term on the right side of the equation as follows

$$\frac{d^2y(t)}{dt^2} = -\frac{K}{m}\left(y + \frac{mg}{K}\right) \tag{7.28a}$$

$$\eta = y + \frac{mg}{K} \tag{7.28b}$$

The derivatives are

$$d\eta = dy \text{ and } \frac{d^2\eta}{dt^2} = \frac{d^2y}{dt^2} \tag{7.29}$$

Using the change of variables in equations 7.28 and 7.29, we can rewrite equation 7.27 as

$$\frac{d^2\eta}{dt^2} = -\frac{K}{m}\eta \tag{7.30}$$

We know very well the solution to 7.30 which is

$$\eta(t) = A \cdot \cos(\omega_o t) + B \cdot \sin(\omega_o t) \tag{7.31}$$

Where $\omega_o = \sqrt{K/m}$. At time 0, we already know that $y(0) = -y_o$ and hence

$$\eta(0) = A = -y_o + \frac{mg}{K} \tag{7.32}$$

We also know that the speed at time t=0 is also zero.

$$u(t) = \frac{d\eta(t)}{dt} = -A \cdot \omega_o \cdot \sin(\omega_o t) + B \cdot \omega_o \cdot \cos(\omega_o t) \qquad (7.33a)$$

and

$$u(0) = 0 = B \cdot \omega_o \rightarrow B = 0 \qquad (7.33b)$$

Using the value for A in equation 7.31 and changing the variable back to y, the solution for the simple harmonic when considering the effect of gravity is

$$y(t) + \frac{mg}{K} = -\left(-y_o + \frac{mg}{K}\right)\cos(\omega_o t) \qquad (7.34)$$

Or

$$y(t) = \left(-y_o + \frac{mg}{K}\right)\cos(\omega_o t) - \frac{mg}{K} \qquad (7.35)$$

This expression is visualized in figure 7.5.

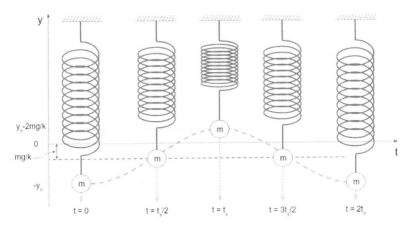

Figure 7.5: Simple harmonic oscillation of the spring when considering gravity.

As seen in the graph the center of oscillation is now shifted by a distance of mg/K below the zero-displacement point. We can obtain this by evaluating the displacement at a time $t = t_\pi/2$ or $\omega_o t_\pi/2 = \pi/2$. Then equation 7.35 gives $y(t_\pi/2) = -\frac{mg}{K}$. That is the same result we found in equation 3.10 in chapter three when demonstrating the use of spring as a scale to measuring weight. Also, the amplitude of the oscillation is reduced to $\left(y_o - \frac{mg}{K}\right)$ instead of y_o when gravity was not considered. The amplitude of the oscillation or the strength of the oscillation is the maximum amount of displacement the spring reaches around the center of the oscillation, which is the factor that is multiplied by the cosine term in equation 7.35. Here, the strength equals the absolute value of that factor. So, gravity has two main effects on the oscillation of the simple harmonic oscillator:

1.   It shifts the center of oscillation to $-mg/K$.
2.   It reduces the amplitude of the oscillation by the same amount, $y_o - mg/K$.

Effect of damping force

Damping force in any spring is a force that tends to slow down the movement. It is always pointing in the opposite direction of the velocity, and it is proportional to it. This damping force helps the spring to return to the zero displacement positions, or as commonly referred to as the equilibrium position. If we neglect gravity at this point, we can write the force equation as

$$m\bar{a} = -Ky\hat{y} - \xi\bar{u} \tag{7.36}$$

Here, ξ is the damping constant that relates friction to velocity. The minus sign indicates that friction is in the opposite direction to the velocity. We know that the velocity vector is along the y axis, $\bar{u} = u\,\hat{y}$, hence the acceleration is as well along the y axis, $\bar{a} = a\hat{y}$. So, we can write equation 7.36 in the scalar form replacing acceleration and speed with their time derivative of y as follows

$$m\frac{d^2y}{dt^2} = -ky - \xi\frac{dy}{dt} \tag{7.37}$$

This is a second order differential equation. To find a solution, let us propose the exponential function as we did in the simple harmonic oscillations before,

$$y(t) = A \cdot e^{\psi t} \tag{7.38}$$

Substituting this solution in equation 7.37 we obtain the following

$$m\psi^2 \cdot Ae^{\psi t} = -K \cdot Ae^{\psi t} - \xi\psi \cdot Ae^{\psi t} \rightarrow m\psi^2 = -K - \xi\psi \tag{7.39}$$

Equation 7.39 can be reduced to

$$\psi^2 + \frac{\xi}{m}\psi + \frac{K}{m} = 0 \tag{7.40}$$

That is a quadratic equation, and the roots of this equation provide the following solutions

$$\psi = -\frac{\xi}{2m} \pm \sqrt{\left(\frac{\xi}{2m}\right)^2 - \omega_o^2} \tag{7.41}$$

Here, $\omega_o = \sqrt{K/m}$. The square root in equation 7.41 contains a subtraction of two terms: $(\xi/2m)^2$ and ω_o^2. That creates three possible scenarios: $\xi/2m < \omega_o$, $\xi/2m = \omega_o$ and $\xi/2m > \omega_o$. The first case, when $\xi/2m < \omega_o$, is commonly referred to as underdamping oscillation. The second case, $\xi/2m = \omega_o$, is the critical damping. The final case is the over-damping motion.

1. Underdamping

That is the case when $\xi/2m < \omega_o$. This makes the subtraction inside the square root to return a negative quantity, $-|\omega_o^2 - \gamma^2|$, where $\gamma = \frac{\xi}{2m}$ and we will refer to it as the damping factor. The square root of this negative quantity is imaginary. Hence, we can write the solution in equation 7.41 as

$$\psi = -\gamma \pm i\sqrt{\omega_o^2 - \gamma^2} \tag{7.42}$$

For simplicity, let us define the square root term in the equation above as $\omega = \sqrt{\omega_o^2 - \gamma^2}$.

We will soon know why we picked this symbol. For now, let's take it as it is. Using this definition, equation 7.42 is simplified to

$$\psi = -\gamma \pm i\omega \tag{7.43}$$

As we did with the simple harmonic oscillations when two solutions are obtained, we consider the summation of both solution for the displacement function as follows

$$y(t) = A \cdot e^{(-\gamma + i\omega)} + B \cdot e^{(-\gamma - i\omega)} \tag{7.44}$$

We apply the initial condition that the initial speed is zero, $u(0) = 0$, or

$$\frac{dy(0)}{dt} = 0 = A \cdot (-\gamma + i\omega) + B \cdot (-\gamma - i\omega) \tag{7.45a}$$

Then

$$\frac{B}{A} = \frac{-\gamma + i\omega}{\gamma + i\omega} \tag{7.45b}$$

Substituting the relation between B and A into equation 1.44, we obtain

$$y(t) = Ae^{-\gamma t} \cdot \left(e^{i\omega t} + \frac{-\gamma + i\omega}{\gamma + i\omega} e^{-i\omega t} \right) \tag{7.46}$$

The other initial condition that we know is the displacement, $y(0) = -y_o$, then

$$y(0) = -y_o = Ae^{-\gamma t} \cdot \left(1 + \frac{-\gamma + i\omega}{\gamma + i\omega} \right) \tag{7.47a}$$

Or

$$A = -y_o \cdot \left(\frac{\gamma + i\omega}{2i\omega} \right) \tag{7.47b}$$

Substituting into equation 7.46

$$y(t) = -y_o e^{-\gamma t} \cdot \left(\frac{\gamma + i\omega}{2i\omega} e^{i\omega t} + \frac{-\gamma + i\omega}{2i\omega} e^{-i\omega t} \right) \tag{7.48}$$

We can rearrange the terms in equation 7.48 as

$$y(t) = -y_o e^{-\gamma t} \cdot \left(\frac{e^{i\omega t} + e^{-i\omega t}}{2} + \frac{\gamma}{\omega} \frac{e^{i\omega t} - e^{-i\omega t}}{2i} \right) \tag{7.49}$$

We know by now that the firm term $\frac{1}{2}(e^{i\omega t} + e^{i\omega t})$ is nothing but $cos(\omega t)$ and the second term $\frac{1}{2i}(e^{i\omega t} - e^{-i\omega t})$ is $sin(\omega t)$. Using these definitions, we can simplify equation 7.49 as

$$y(t) = -y_o e^{-\gamma t} \left(\cos(\omega t) + \frac{\gamma}{\omega} \sin(\omega t) \right) \tag{7.50}$$

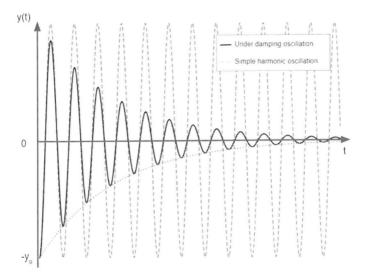

Figure 7.6: Underdamping oscillation compared with simple harmonic oscillations.

This is an interesting result as it gives us a response that is similar to simple harmonic oscillation except for a few differences. First, the frequency of oscillation is less than the natural frequency, $\omega = \sqrt{\omega_o^2 - \gamma^2}$. Also, the amplitude of oscillation is exponentially decaying in time, $-y_o \cdot e^{-\gamma t}$. This motion is commonly referred to as underdamping oscillation as visualized in figure 7.6. One can observe clearly that the underdamping oscillation in equation 7.50 becomes the same as that of the simple harmonic oscillation in equation 7.12 when setting the damping factor to become zero, $\gamma = 0$.

2. Critical damping

If we increase the damping coefficient (or decrease the natural frequency) such that it equates the natural frequency, the case when $\gamma = \xi/2m = \omega_o$, then the term inside the square root vanishes and the damping constant becomes $\xi = 2\omega_o m$. In this case, the frequency approaches zero value or $\omega \to 0$. This would seem to create a problem in our previous solution in 1.50 as ω is present at the denominator in the second term, $\frac{\gamma}{\omega} sin(\omega t)$. This would for sure be a problem if it was not for the fact that the sinusoidal term as well goes to zero when $\omega \to 0$. That gives us a division of zero by zero. To find the value of this term we wound need to apply L'Hopital's rule as

$$\lim_{\omega \to 0} \gamma \frac{\sin(\omega t)}{\omega} = \lim_{\omega \to 0} \gamma \frac{d(\sin(\omega t))/d\omega}{d(\omega)/d\omega} = \lim_{\omega \to 0} \gamma t \cos(\omega t) = \gamma t \qquad (7.51)$$

Using this result, we can now write the critical damping motion as

$$y(t) = -y_o e^{-\gamma t}(1 + \gamma t) \qquad (7.52)$$

At the critical damping condition, the friction force in the system allows the stretched spring to return gradually to the zero position, commonly referred to as the equilibrium point, without going through oscillations. A common example in our daily life is the shock absorber placed in the car wheels. Its job is exactly that it absorbs the stretching or pressing on the wheels and let it go back gradually, preferably as fast as possible, to the equilibrium position. The critical damping motion is visualized in figure 1.7. In figure 1.7, we plotted the damping oscillation condition while reducing the natural frequency, by reducing the spring constant, to approach the damping coefficient reaching the critical damping condition.

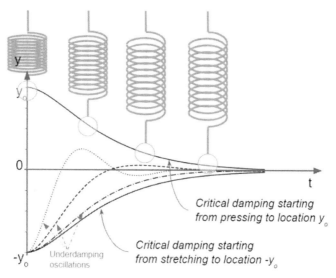

Figure 7.7: Critically damped motion for two initial conditions, upper case is when the spring is pressed to location y_0, and the lower curve is when the spring was stretched to initial location $-y_0$. The dotted curves show the underdamped oscillations for the purpose of comparison.

3. Over damping

The last scenario here is when $\gamma > \omega_o$. In this case, the term inside the square root in equation 7.42 is now positive.

$$\psi = -\gamma \pm \sqrt{\gamma^2 - \omega_o^2} \tag{7.53}$$

For simplicity let us define the square term as $\eta = \sqrt{\gamma^2 - \omega_o^2}$. So, as we have two solutions, $\psi = -\gamma + \eta$ and $\psi = -\gamma - \eta$, then the displacement function is the combination of both as follows.

$$y(t) = A \cdot e^{-\gamma t + \eta t} + B \cdot e^{-\gamma t - \eta t} \tag{7.54}$$

Applying the first condition that the initial velocity is 0, $u(0) = \frac{dy(0)}{dt} = 0$, then

$$\frac{dy(0)}{dt} = 0 = A \cdot (-\gamma + \eta) + B \cdot (-\gamma - \eta) \tag{7.55a}$$

$$\frac{B}{A} = \frac{-\gamma + \eta}{\gamma + \eta} \tag{7.55b}$$

Substituting in equation 7.54

$$y(t) = Ae^{-\gamma t} \cdot \left(e^{\eta t} + \frac{-\gamma + \eta}{\gamma + \eta} \cdot e^{-\eta t}\right) \tag{7.56a}$$

$$y(t) = \frac{Ae^{-\gamma t}}{\gamma + \eta} \cdot ((\gamma + \eta) \cdot e^{\eta t} + (-\gamma + \eta) \cdot e^{-\eta t}) \tag{7.56b}$$

$$y(t) = \frac{Ae^{-\gamma t}}{\gamma + \eta} \cdot (\eta \cdot (e^{\eta t} + e^{-\eta t}) + \gamma \cdot (e^{\eta t} - e^{-\eta t})) \tag{7.56c}$$

The first square bracket is twice a hyperbolic cosine or cosh function in short. The second square bracket is twice hyperbolic sine or sinh function in short.

Hyperbolic cosine:	$\cosh(a) = \frac{1}{2}(e^a + e^{-a})$
Hyperbolic sine:	$\sinh(a) = \frac{1}{2}(e^a - e^{-a})$

The displacement function becomes

$$y(t) = \frac{2Ae^{-\gamma t}}{\gamma + \eta} \left(\eta \cdot cosh(\eta t) + \gamma \cdot sinh(\eta t) \right) \tag{7.57}$$

The initial displacement is $y(0) = -yo$, then

$$y(0) = -y_o = \frac{2A\eta}{\gamma + \eta} \rightarrow A = -\frac{1}{2} y_o (1 + \gamma/\eta) \tag{7.58}$$

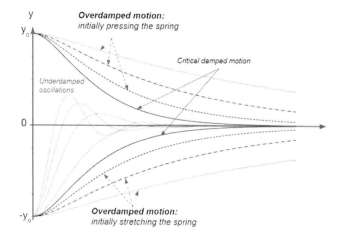

Figure 7.8: Overdamped motion for the case of initially pressed or stretched spring compared to critical damping and underdamping motions.

Using this result, we obtain

$$y(t) = -y_o e^{-\gamma t} \left(\cosh(\eta t) + \frac{\gamma}{\eta} \cdot \sinh(\eta t) \right) \tag{1.59}$$

Figure 7.8 visualizes the three types of damping motions, underdamped in dashed blue line, critical damping in solid line and overdamping in black dashed lines. For overdamped motions, the spring moves towards the equilibrium point at a slower rate compared to critical damping. In both critical and overdamped motions, the spring does not oscillate; it rather follows a gradual trajectory

Table 7.2: Summary of spring motion and their conditions.

Motion type	Condition	Displacement
None-damping (simple harmonic oscillation)	Damping coefficient = 0, $\gamma = 0$	$y(t) = y(0) \cdot \cos{(\omega_o t)}$, $\omega_o = \sqrt{K/m}$ K is the spring constant m is the attached mass $y(0)$ is the initial displacement
Underdamped oscillation	$\gamma < \omega_o$	$y(t) = y(0)e^{-\gamma t} \cdot (\cos(\omega t) - \frac{\gamma}{\omega} \cdot \sin(\omega t))$ Damping coefficient $\gamma = \frac{\xi}{\omega}$ Spring damping constant: ξ Oscillation frequency: $\omega = \sqrt{\omega_o^2 - \gamma^2}$
Critical damping motion	$\gamma = \omega_o$	$y(t) = y(0)e^{-\gamma t} \cdot (1 + \gamma t)$
Overdamping motion	$\gamma > \omega_o$	$y(t) = y(0)e^{-\gamma t}(\cosh(\eta t) + \frac{\gamma}{\eta}\sinh(\eta t))$ $\eta = \sqrt{\gamma^2 - \omega_o^2}$

Pendulum motion

One of the common mechanical oscillators is the pendulum shown in figure 7.9. It consists of a mass m that is attached to a rod that has a length R and is by its turn attached to a joint on the ceiling. This allows the rod to swing along the θ direction. If we lift the mass to an initial height L then we let it go as in the figure, then the only force that acts on it is the gravitational force.

Here, only the θ component of the gravitational force affects the motion of the mass. This is because the tension in the rod cancels out the radial component of the force and hence the acceleration is zero along the radial direction. We can then write the following equation of motion.

$$F_\theta = mg \sin \theta = ma_\theta \tag{7.60}$$

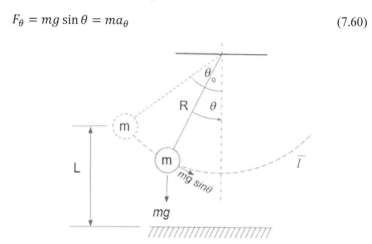

Figure 7.9: Pendulum motion along the θ direction

The acceleration is the second derivative of the displacement along the motion path, l, with respect to time, $a_\theta = \frac{d^2 l}{dt^2}$. The circular path length at any angle θ starting from an initial angle θ is $l = R(\theta - \theta_o)$.

$$a_\theta = \frac{d^2}{dt^2}(\theta_o - \theta) = -R\frac{d^2\theta}{dt^2} \tag{7.61}$$

We can re- write the force equation as

$$g \sin \theta = -R\frac{d^2\theta}{dt^2} \tag{7.62}$$

Equation 7.62 might look simple at the first glance, however finding a closed form solution for such an equation is not an easy task. When facing such a difficulty, we could look for specific cases where a simplification could be applied or use discrete methods and solve the problem on the computer using numerical computations. For now, let us consider a simplified case when the range of the angles that the pendulum swings is very small, $\theta_o \ll 2\pi$. We can now use Taylor's expansion of the sinusoidal function.

$$\sin \theta = \theta - \frac{1}{3!}\theta^3 + \frac{1}{5!}\theta^5 - \frac{1}{7!}\theta^7 + \ldots \quad (7.63)$$

When the angles are very small, then we can neglect the higher powers of θ and the sinusoidal function is simplified to

$$\sin \theta \approx \theta \quad (7.64)$$

This simplifies equation 7.62 to

$$g\theta = -R \frac{d^2\theta}{dt^2} \quad (7.65)$$

This is rather similar to the simple harmonic oscillation and we know that the solution should be in the following form

$$\theta(t) = \theta_o \cos(\omega t) \quad (7.66)$$

This again forms simple harmonic oscillation with an angular frequency of $\omega = \sqrt{g/R}$.

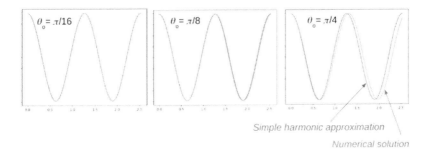

Figure 7.10: Simplified simple harmonic solution (green) compared to numerical solution (orange) for the pendulum motion.

Figure 7.10 compares the simple harmonic oscillation approximation in equation 7.66 to numerical solution to equation 7.62. One can notice that for small angle (in the example above $\theta_o \leq \pi/16$) the two solutions are almost identical. This is however not the case when the angle gets larger as the difference is clearly seen for $\theta_o = \pi/4$. Hence, for most of the practical use of the pendulum, the simple harmonic approximation can provide a sufficient means to describe its oscillating motion.

7C. Electrical oscillators

So far, we have talked about oscillations in a mechanical system where the motion of a mass in a system follows a form of simple or damping oscillations. When thinking about electrical oscillators one expects electrical charges to experience an oscillating motion, which would result in a measurable quantity that can be mathematically expressed in a similar form to simple harmonic and damping oscillations. However, before going into finding an example of such oscillators, one would need to introduce some basic electrical components and quantities that we need to familiarize ourselves with.

Review of some basic electrical components

1. Electrical charges and current

Electrical charge is a property of the matter which defines how they are affected by electric or magnetic sources. These charges can be either positive such as proton or negative such as electrons. The unit of charge is Coulomb or C for short. We usually use the symbol q for elementary charges and Q for the total charge in the system. When charges move, then for a fixed volume in space, the total charge Q will vary with time. This variation of charge with time is defined as the electrical current and it has units of Ampere or A for short.

$$I = \frac{dQ}{dt} \tag{7.67}$$

Notice that the definition of 1 C is the amount of charge that passes through a certain area due a current of one Ampere over a time interval of one second.

2. Electromotive force

To generate charges and electrical current an electrical source is needed. One common example of such sources is a battery. In batteries, the chemical reactions inside of them generate an energy that moves charges from the battery to an electrical circuit that is connected to it. The amount of work (the energy consumed) that is needed to move a unit charge (1 C) is the electromotive force, emf, and it is measured in volts. Typically, batteries are made to provide a constant amount of emf between its two ports (positive and negative). Example, AAA batteries provide emf of 1.5 V.

$$emf = \frac{W}{Q} \text{(units of } J/C = Volts\text{)} \tag{7.68}$$

3. Electrical resistance

When the battery is connected to an electrical circuit, charges start to move in the circuit. However, part of these charges dissipates in the form of heat due to the opposition of the material and objects in the circuit to the flow. This loss mechanism is typically represented by a quantity known as resistance, R, and it is measured in Ohm or Ω.

4. Electrical potential difference

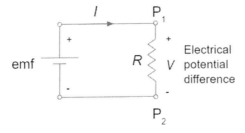

Figure 7.11: The generated emf by the source (battery) builds a potential difference across the resistor and causes changes to the flow of charges (current) through the circuit.

If we connect the battery to a resistance R, as in figure 7.11, then it will perform a work to move charges to points P_1 and P_2 as in the figure (positive charges to P_1 and negative charges to P_2). This work performed in moving the charge from a reference (the source) to a point (P_1 or P_2) is defined as the electrical potential at that point. The difference between the electrical potential at two different points is the electrical potential difference, V, between them. Hence, as in figure 7.11, the electrical potential difference across the resistance is the electrical potential difference between P_1 and P_2.

If the resistance is zero, then zero potential would be needed to bring the charges from the source as current would flow naturally with no opposition. Increasing the value of resistance would require increasing the work needed and hence the potential difference to make up for the loss and sustain a constant current flow. Increasing the resistance further requires higher potential to ensure the flow of the same amount of current. This describes a linear relationship between the potential difference V and the resistance

$$V = IR \tag{7.69}$$

This is commonly known as Ohm's law.

5. Capacitor

A capacitor is a device that has the capability of storing electrical charges. In the most common form, it is made of two metallic parallel plates separated by a non-conducting medium (such as air or plastic) as shown at the right side of the circuit in figure 7.12. In this circuit a capacitor (two plates) is connected in parallel to a resistance R. The switch when pressed would connect the two components to a battery with a fixed emf value. If we pressed the switch, then the charges move in the circuit and a potential difference V is built across the capacitor. Due to the non-conducting gap of separation d in the capacitor, an electric field is generated inside the capacitor. The unit of the electric field is V/m. For now, we can think of the electric field as the force per unit charge that holds the charges on the capacitor plates. We know that the electrical potential is the work performed to bring the charges to the two sides of the capacitors. Hence, the electric field is expected to be proportional to the potential difference across the capacitor.

$$E = V/d \tag{7.70}$$

We will learn more about the electric field later in the electromagnetics chapter. For now, we accept our earlier statement that this field exerts force on the charges making them pile at the two plates of the capacitor: positive charges on top and negative at the bottom as in the figure. Hence, we could define the capacitance of the capacitor as the number of charges stored due to an electrical potential difference of one volt, or

$$C = Q/V \tag{1.71}$$

The unit of capacitance is Farad or F in short. 1 Farad stands for the number of charges stored due to 1 V of electrical potential difference.

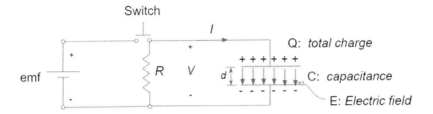

Figure 7.12: Capacitor with an electrical potential V across it due to the accumulation of a total charge Q.

Damping/discharging effect

Now, let us release the switch so there is no more electromotive force that pushes the charges in the circuit. If there was no resistance connected to this capacitor, then in a perfect world the charge Q should always remain on the capacitor as there is no source of loss. However, when the resistor R is present with a finite value the charges start to move in the wire and a current flow from the capacitor to the resistance as shown in the initial state in figure 7.13. This causes that part of the electrical potential energy to dissipate in other forms, such as heat, inside the resistor. That causes a reduction in the current and hence the potential difference across the resistance, $V_R = IR$, as well.

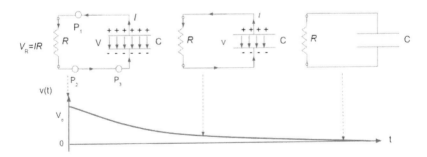

Figure 7.13: Draining of an originally charged capacitor to a resistance R.

Consider the three points in figure 7.13 P_1, P_2 and P_3. The potential difference between points P_2 and P_3 should equal the addition of the potential differences between P_1, P_2 and P_1,P_3, or

$$V_{23} = V_{12} + V_{23} \tag{7.72}$$

We could directly identify that $V_{13} = V$, which is the potential difference across the capacitance. Also, $V_{12} = I\,R$, which is the potential difference across the resistance. However, it is evident that $V23 = 0$ as points P$_2$ and P$_3$ are practically the same point. Hence, we could write that

$$V + IR = 0 \tag{7.73}$$

This is known as Kirchhoff's voltage law.

Kirchhoff's voltage law statement

The sum of the potential differences (voltages) around any closed loop is zero.

Let us rearrange equation 7.71 as

$$CV = Q \tag{7.74}$$

If we take the time derivative of equation 7.74, then we could find that

$$C\frac{dV}{dt} = \frac{dQ}{dt} \tag{7.75}$$

The rate of change of the charges per unit time is the current as stated earlier, $I = \frac{dQ}{dt}$. Hence, equation 7.75 becomes

$$I = C\frac{dV}{dt} \tag{7.76}$$

From equation 7.73 we know that $I = -V/R$, hence we can rewrite equation 1.76 as

$$V = -RC\frac{dV}{dt} \tag{7.77}$$

The solution of equation 7.77 is an exponentially decaying function

$$V(t) = V_o \cdot e^{-\left(\frac{t}{RC}\right)} \tag{7.78}$$

Where V_o is the initial potential difference across the capacitance. This function represents a discharging capacitance as shown at the bottom of figure 7.13. The initial potential $V_o = Q_o/C$, where Q_o is the initial charge stored in the capacitor at time $t = 0$.

Generating an electrical oscillator

The scenario explained above considered that the resistance dissipated the electrical potential energy (in the form of heat). Now, consider that we replace the resistance with a component that instead of dissipating the energy, it stores it in a different form as shown hypothetically in figure 7.14a.

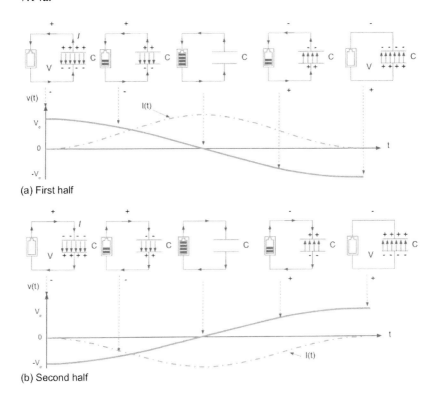

Figure 7.14: Cycle of oscillation for an electrical oscillator of a capacitor and a load that stores the energy.

We can break the process as follows:

1. When the capacitor is fully charged at the initial stage, the current flows from the positive charges towards the negative plate.

2.   The introduced component is charged by the flow of the current through it.

3.   The component reaches a maximum capacity when the capacitor is fully discharged as in the middle of figure 7.14a.

4.   At that point the potential drop across the capacitor is zero while the current is at maximum.

5.   After this point the capacitor will start to build positive chargers on the bottom plate due to the current flow and the component starts to discharge.

6.   The potential starts to build across the capacitor however with an opposite sign compared to the original state.

7.   The capacitor is fully charged when the current drops to zero while the potential difference reaches a maximum amplitude $-V_o$ as shown at the right of figure 7.14a. That is the opposite sign of the original electric potential as shown in figure 7.14a.

8.   After this point, a current starts to flow from the positive plate towards the negative one in a direction that is opposite to the original state.

9.   The capacitor discharges into the component till a maximum current with opposite direction is reached as shown in the middle of figure 7.14b.

10.   The component then starts to charge the capacitor till a maximum potential of V_o is achieved returning the system to the initial state.

11.   This process then repeats.

This process has a good similarity to the spring mechanical oscillator in figure 7.4. The energy was oscillating between potential energy stored in the spring and as kinetic energy of the moving mass. By analogy, in this hypothetical electrical oscillator, potential energy is stored in a form of static charges around the capacitor that builds a potential difference V. The kinetic energy generated by the moving charges is stored in the introduced element. Accordingly, maximum potential energy is obtained when the electrical potential difference is maximized, while maximum kinetic energy is reached when the current is maximum. A known component that is capable of storing current in a different form and generates an electromotive force is the inductor.

Inductor

An inductor or coil is a device that opposes the flow of the electrical current through it. In comparison to the resistance, the current generates a magnetic field, B, around it as illustrated in figure 7.15. We could think of the magnetic field as the force per unit charge that moves with a speed of 1 m/s. It has units of A/m. A is the short of Ampere which is the unit of current as explained earlier. In comparison to the electric field which points from the positive static charge to the negative one, the direction of the magnetic field lines follows the right-hand rule relative to the current as demonstrated at the top of figure 7.15. The summation of all the magnetic field lines that moves through the coil area is known as magnetic flux, ϕ, and it has units of Weber or Wb for short. The amplitude of the magnetic flux produced by one Ampere of current is defined as the inductance of the coil, L. Hence,

$$L = \frac{\phi}{I} \qquad\qquad (7.79)$$

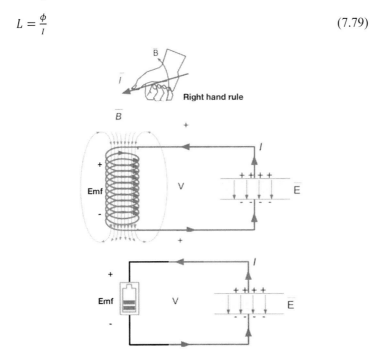

Figure 7.15: The inductor and capacitor circuit electrical oscillator where the change in the magnetic field B causes an electric field that charges the capacitor. The field causes the charges to move to create a current that contributes to the magnetic field.

This looks like Ohm's, $R = V/I$. However instead of dissipating electrical energy into heat, it generates a magnetic flux. The term L stands for inductance, and it has units of Henry or H for short. According to Faraday's law of induction, the time variation of the magnetic field generates an electromotive force

$$emf = -\frac{d\phi}{dt} \qquad (7.80)$$

In other words, the variation of the magnetic flux generated a source (storing energy that functions like a battery). The negative sign indicates that the emf opposes the increase in the flux. This is shown in figure 7.15 where the coil is replaced by an energy charging element with an emf that has a positive sign on the top. That generates a current with the opposite direction of the input current. The potential difference across the inductor has a negative sign to emf, $V_L = -emf$.

From equation 7.79, we know that the flux is proportional to the current flow. If the inductance is assumed not to vary with time, we can re-write equation 7.80 as

$$emf = -L\frac{dI}{dt} = -V_L \qquad (7.81)$$

If we neglect the resistance in the circuit for now and following Kirchhoff's voltage law, one could say that $V + V_L = 0$ and hence

$$V = -L\frac{dI}{dt} \qquad (7.82)$$

We know that the total current flowing through the circuit, $I = CdV/dt$. Hence,

$$V = -LC\frac{d^2V}{dt^2} \qquad (7.83)$$

Now, here is the result we were waiting for and had to go through all the trouble to reach. If we look carefully at the expression $\frac{d^2V}{dt^2} = -\frac{1}{LC}V$ in equation 7.83, we find it strikingly similar to equation 7.7. We know the solution should follow a simple harmonic oscillation in the form

$$V(t) = V_o \cdot cos\left(\frac{t}{\sqrt{LC}}\right) \qquad (7.84)$$

Where the natural frequency of the oscillator is $\omega_o = \frac{1}{\sqrt{LC}}$. This is exactly what we were after. The potential difference across the capacitance (as well as the current going through the circuit) follows a simple harmonic oscillation form when the resistance is neglected in the circuit.

Resistance and damping oscillations

The circuit diagram in figure 1.14 shows the case when resistance is not neglected. Following Kirchhoff's voltage law for the circuit in figure 7.16, we find that

$$V_L + V_R + V = 0 \rightarrow V = -V_L - V_R \qquad (1.85)$$

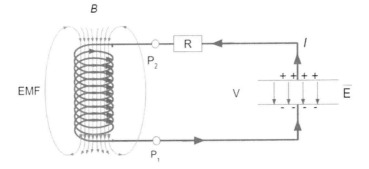

Figure 7.13: Damping electrical oscillator due to the presence of the resistance R in the circuit.

Substituting the relation between V_L and V_R in terms of current into equation 7.85 we obtain

$$V = -LC\frac{d^2V}{dt^2} - RC\frac{dV}{dt} \qquad (7.86)$$

Dividing both sides by LC and moving the right-hand side to the left we obtain

$$\frac{d^2V}{dt^2} + \frac{R}{L}\frac{dV}{dt} + \frac{1}{LC}V = 0 \qquad (7.87)$$

Equation 7.87 has a strong resemblance to equation 7.37 where the natural frequency here is $\omega_o = \frac{1}{\sqrt{LC}}$ and the damping factor is $\gamma = \frac{R}{2L}$. We could right away, predict the different ways the potential drop over the capacitor, V(t), varies in time depending on the relation between γ and ω_o as we did for the damping motion in section one. We could use the results obtained from before to build table 7.3.

Table 7.2: Summary of spring motion and their conditions.

Motion type	Condition	Displacement
Non-damping (simple harmonic oscillation)	Damping coefficient = $0, \gamma = 0$	$V(t) = V(0) \cdot cos(\omega_o t)$, $\omega_o = 1/\sqrt{LC}$ L is the Inductance C is the capacitance $V(0)$ is the initial potential difference
Underdamped oscillation	$\gamma < \omega_o$	$V(t) = V(0)e^{-\gamma t} \cdot (cos(\omega t) + \frac{\gamma}{\omega} \cdot sin(\omega t))$ Damping coefficient $\gamma = \frac{R}{2L}$ R is the resistance. Oscillation frequency: $\omega = \sqrt{\omega_o^2 - \gamma^2}$
Critical damping motion	$\gamma = \omega_o$	$V(t) = V(0)e^{-\gamma t} \cdot (1 + \gamma t)$
Overdamping motion	$\gamma > \omega_o$	$V(t) = V(0)e^{-\gamma t}(cosh(\eta t) + \frac{\gamma}{\eta} sinh(\eta t))$ $\eta = \sqrt{\gamma^2 - \omega_o^2}$

Example 7.1

Determine the type of oscillations and for the following R, C, L values:

1- L=10 nH, C = 100 pF, R=0 Ω
2- L=10 nH, C= 100 pF, R=100 Ω
3- L=100 nH, C= 10 pF, R=100 Ω
4- 100 nH, C= 10 pF, R=200 Ω

Solution:

1- $\gamma = R/2L = 0$, $\omega_o = 1/\sqrt{10 \times 10^{-9} \times 100 \times 10^{-12}} = 10^9 rad/s$.

$\gamma = 0 \rightarrow$ Simple harmonic oscillation.

2- $\gamma = R/2L = 100/2 \times 10 \times 10^{-9} = 5 \times 10^9$, $\omega_o = 10^9$.

$\omega_o < \gamma \rightarrow$ Over-damping motion.

3- $\gamma = 100/2 \times 100 \times 10^{-9} = 5 \times 10^8$, $\omega_o = 1/$
$\sqrt{10 \times 100^{-9} \times 10 \times 10^{-12}} = 10^9 rad/s$. $\omega_o > \gamma \rightarrow$ Under-damping oscillation.

4- $\gamma = 200/2 \times 100 \times 10^{-9} = 10^9$, $\omega_o = 1/$
$\sqrt{10 \times 100^{-9} \times 10 \times 10^{-12}} = 10^9 rad/s$. $\omega_o = \gamma \rightarrow$ Critical-damping motion.

7D. Summary

- Simple harmonic oscillation
  - Neglecting gravity
    - $\frac{d^2y(t)}{dt^2} = -\omega_o^2 \cdot y(t)$
      - Natural frequency, ω_o
      - $\omega_o = \sqrt{k/m}$
      - $f = \frac{1}{T} = \omega_o/2\pi$
      - $\phi(t) = \omega_o \cdot t$
    - $y(t) = -y_o \cdot \cos(\omega_o \cdot t)$
  - System energy
    - Potential energy
      - $PE = \int_0^y Kydy = \frac{1}{2}Ky(t)^2$
      - $PE = \frac{1}{2}Ky_o^2\cos^2(\omega_o t)$
    - Kinetic energy
      - $KE = \frac{1}{2}mu(t)^2$
      - $KE = \frac{1}{2}my_o^2\omega_o^2 \cdot \sin^2(\omega_o t)$
    - Total energy
      - $E = \frac{1}{2}Ky_o^2$
      - System energy is always constant
  - Effect of gravity
    - $\frac{d^2y(t)}{dt^2} = -\omega_o^2\left(y + \frac{mg}{K}\right)$
    - $y(t) = \left(-y_o + \frac{mg}{K}\right)\cos(\omega_o t) - \frac{mg}{K}$
    - Center of oscillation
      - $-mg/K$
    - Amplitude of oscillation
      - $y_o - mg/K$
  - Effect of damping force
    - $\frac{d^2y}{dt^2} = -\omega_o^2 y - \frac{\xi}{m}\frac{dy}{dt}$
    - Solution: $y(t) = A \cdot e^{\psi t}$
      - $\psi = -\gamma \pm \sqrt{\gamma^2 - \omega_o^2}$
      - $\gamma = \frac{\xi}{2m}$

- Underdamping
  - $\xi/2m < \omega_o$
  - $\psi = -\gamma \pm i\omega$
  - $\omega = \sqrt{\omega_o^2 - \gamma^2}$
  - $y(t) = -y_o e^{-\gamma t}\left(\cos(\omega t) + \frac{\gamma}{\omega}\sin(\omega t)\right)$
- Critical damping
  - $\gamma = \xi/2m = \omega_o$
  - $\omega \to 0$
  - $y(t) = -y_o e^{-\gamma t}(1 + \gamma t)$
- Overdamping
  - $\psi = -\gamma \pm \eta$
  - $\eta = \sqrt{\gamma^2 - \omega_o^2}$
  - $y(t) = -y_o e^{-\gamma t}\left(\cosh(\eta t) + \frac{\gamma}{\eta} \cdot \sinh(\eta t)\right)$
- Pendulum motion
  - $F_\theta = mg \sin\theta = ma_\theta$
  - $g \sin\theta = -R\frac{d^2\theta}{dt^2}$
  - Simplified solution, small angles
    - $\sin\theta \approx \theta$
    - $\theta(t) = \theta_o \cos(\omega t)$
- Electrical oscillators
  - Electrical current
    - Change of the electrical charge per unit time
    - $I = \frac{dQ}{dt}$
  - Electromotive force
    - Work to move one unit charge
    - $emf = \frac{W}{Q}$ (units of $J/C = Volts$)
  - Electrical resistance
    - The loss of electrical charges in form of heat.
    - The loss is defined as resistance, R (Ohm or Ω)
  - Electrical potential
    - The work needed to move a unit charge from a reference to one point

- Electrical potential difference
  - The difference of the electrical potential between two points
  - Symbol, V, and units of Volts.
  - In terms of resistance
    - $V = IR$
- Capacitor
  - Stores a charge Q when a potential difference V is applied
    - $C = Q/V$
  - Electrical field between the capacitor plates is
    - $E = V/d$
- Inductor
  - Coil that opposes the flow of electric current through
  - Magnetic flux, ϕ
    - The summation of all magnetic lines
    - $\phi = LI$
    - Units of Weber or Wb
  - Inductance
    - The magnetic flux that caused by 1 Ampere
    - $L = \dfrac{\phi}{I}$
    - Units of Henry
- Kirchhoff's voltage law
  - The directed sum of the potential differences (voltages) around any closed loop is zero
- Damping effect
  - For an R-C circuit
    - Potential different variation in time
      - $V = -RC\dfrac{dV}{dt}$
      - $V(t) = V_o \cdot e^{-\left(\frac{t}{RC}\right)}$

- Generating electrical oscillator
  - For L-C circuit
    - The potential difference across the capacitor
      - $V = -LC \frac{d^2V}{dt^2}$
      - $V(t) = V_o \cdot \cos\left(\frac{t}{\sqrt{LC}}\right)$
- Damping oscillations
  - For R-L-C circuit
  - $\frac{d^2V}{dt^2} + \frac{R}{L}\frac{dV}{dt} + \frac{1}{LC}V = 0$

7E. Review questions

1 – Determine the type of oscillations for the following spring and weight parameters:

Mass (kg)	Spring const K (N/m)	Damping const. (Kg/sec)	Type of oscillation
10	6	0.5	
10	1	8	
5	1	0	
10	2.5	10	

2 – Design a simple harmonic oscillator that produces a natural frequency of 20 Hz.

3 – What is the length of the pendulum needed to produce an angular frequency of 40 rad/sec?

4 – When four people with a combined mass of 325 kg sit down in a 2000-kg car, they find that their weight compresses the springs an additional 0.55 cm.

    a)   What is the effective force constant of the springs?

    b)   The four people get out of the car and bounce it up and down. What is the frequency of the car's vibration?

    c)   What would be the needed damping coefficient to remove the bouncing effect?

5 – Design an LC circuit that produces an oscillating frequency of 2 kHz.

6 – For the circuit designed above, what are the resistance values that would cause underdamping and overdamping?

7 – For an LRC circuit with components values as in the table, determine what type of oscillation it produces.

C (F)	L (mH)	R ()	Type of oscillation
0.2	0.3	0	
0.4	0.2	10	
0.4	0.2	100	
0.4	4	100	

8 – For the circuit shown below, the applied emf is 5V. When the switch was closed so the emf is connected to the circuit, a total charge of $Q = 50 \times 10^{-6}$ C was built on the capacitor plate and the maximum current obtained was 200 mA. Calculate the lifetime of the charges on the capacitor when the switch is open. Note that the lifetime is the time the potential difference across the capacitor takes to drop to 10% of its original value.

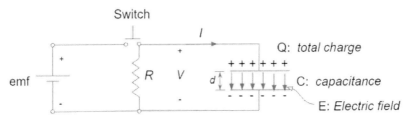

9 – For the circuit above, if 1 mH inductor was added in series with the capacitor, determine the type of oscillation obtained by the circuit.

10 – Design an LRC oscillator circuit that produces radio frequency waves with a frequency of 1 GHz. The circuit should have a damping ratio of 0.7 (damping ratio $\xi = \gamma/\omega_o$). Select component values for the inductor, capacitor, and resistor that will produce the desired frequency and damping characteristics. Explain the process you used to determine the component values and analyze the circuit's performance using a frequency response plot.

CHAPTER EIGHT

WAVES

8A. Introduction to waves

In the previous unit, we explained oscillation and how it represents a motion or a change of a quantity around an equilibrium state. For example, stretching the spring with an attached mass causes the location of the mass to alternate back and forth around the zero-stretch location. The same was achieved when transferring the electromagnetic energy back and forth between the capacitor and the conductor. These oscillations introduce a frequency that represents the number of complete cycles per unit time. The question is, can we transfer these oscillations to neighboring systems or through space? Can these oscillations propagate?

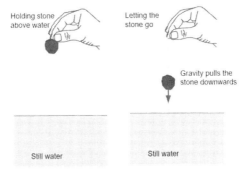

Figure 8.1: The hand holds a stone at a height h from the water surface. It then let it go. The stone gains speed, hence kinetic energy and momentum, due to gravity as it travels downwards before hitting the surface.

A stone in water

To answer the question, let us visit one of our childhood's activities, throwing a stone in still water as shown in figure 8.1. As the stone moves downwards due gravity, it builds up a speed and hence kinetic energy and momentum. Once it hits the surface, part of this kinetic energy is transferred in moving the water surface downwards. The water surface will reach a certain depth before the surface tension (the pulling force by the water surface molecules that tend to keep the water surface at minimum surface area) starts pulling the surface up. The surface moves upwards passing the original level and reaching a peak. The surface tension pulls it down again and the cycle repeats itself as visualized in figure 8.2. This reminds us of the spring underdamping oscillations. The plot at the bottom in figure 8.2

shows the water level at the central location versus time. That clearly shows an oscillation with a period T and frequency $v = 1/T$ Hz.

In contrast to oscillators, the water surface did not experience only fluctuations at the point where the stone fell. These fluctuations affect the neighboring regions, and it starts to propagate away from the central region as shown by the horizontal arrows in figure 8.2.

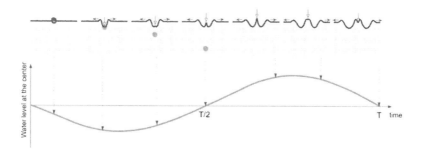

Figure 8.2: When a small stone falls into a still water, the water level at the center moves in an oscillating motion due to the surface tention.

Wavelength

Figure 8.3 shows the evolution of the surface fluctuations through time and space. When the central point oscillates a complete cycle during a period T, the edge of the wave propagates a distance L from the center. Now, as the motion of the central point is of oscillating nature, we expect that every time the oscillation completes a cycle T, the edge moves an extra distance L further away from the center.

If we imagine the shape of the water surface after three complete cycles, or $t = 3T$, one can say that the edge moved by a distance of 3L and inside each cycle the surface has a sinusoidal shape (half dome above the water level and another half below the water level) as depicted in figure 8.4. The surface moves up and down in time and this movement propagates in distance/space. This behavior is commonly known as a wave.

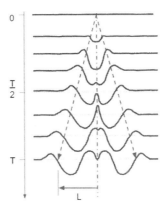

Figure 8.3: The evolution of the surface fluctuations through time and space. When the central point oscillates a complete cycle, the edge of the wave propagates a distance K from the center. As the motion repeats every time the oscillation completes a cycle T, the edge moves an extra distance L further from the center.

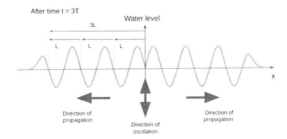

Figure 8.4: Water surface fluctuations after time of 3T from the moment the stone fell in the water.

A wave consists of a repeated pattern in space, cycles that repeat every distance L. This length of the cycle is commonly known as the wavelength, and we commonly use the Greek letter λ to represent it. As it is a length, it has units of meters.

The wave as well propagates and has a direction and speed of propagation. We know that for each complete oscillation of time T, the wave moves by a distance L completing a cycle in space as well. We have just learned that the length of the cycle is referred to as the wavelength and the frequency of oscillation is the inverse of the period T.

Frequency: $v = 1/T (Hz)$ (8.1a)

Wavelength: $\lambda = L (m)$ (8.1b)

As shown in figure 8.4, the wave travels a wavelength every one time-cycle. Hence, the wave travels with a speed that equals the ratio of the wavelength to the period or

$$u = \frac{\lambda}{T} = \lambda v$$ (8.2)

The speed at which the wave propagates is commonly known as the phase velocity. Now a bright student would argue that I have defined a scalar quantity, yet it was called velocity, which is a vector quantity. The answer is that equation 8.2 defines the amplitude of the phase velocity, while the direction of the velocity is along the direction of the wave propagation. The question now is what direction does the wave propagate?

8B. Wave propagation

To understand how and where the waves propagate, one needs to get back to the stone in the water example and examine the view from above at different time instances. What do we see? Most of us already have the answer ready in mind. We see circles (waves) expanding and moving away from the central point where the stone was first thrown as seen in figure 8.5.

Figure 8.5: Top view of the wave propagation at three different cycles. The dark ring represents dip in the water surface while the bright ring represents a peak.

In the figure the dark rings represent the dips in the water surface, when the water is at minimum height. The bright rings represent the peaks in the water surface, when the water is at maximum height. As shown, the peaks and dips are moving away from the central region. In other words, the oscillations of the water surface at the center generates a wave that propagates along the radial direction away from the center.

Mathematical representation

If we mark the water profile along the radial direction at a specific instant of time t, one can depict the plot in figure 8.6. Mathematically we can represent the profile of the water height as a sinusoidal function of the radial direction as

$$h(r,t) = A \cdot \sin\left(\frac{2\pi r}{\lambda} + \phi(t)\right) \qquad (8.3)$$

where $r = \sqrt{x^2 + y^2}$. Here we neglect the damping effect. The constant A is the maximum height the wave reaches. The term $2\pi r$ indicates that the phase increases by 2π, or one cycle, every time the distance increases by a wavelength λ. For example, if we mark a distance $r = 2\lambda$ then the total phase changes by $\frac{2\pi \cdot 2\lambda}{\lambda} = 4\pi = 4\pi$ or two cycles, $2 \cdot 2\pi$. The last term, $\phi(t)$, is a constant phase that is due to the fact that a time t has elapsed from the instant the oscillations started.

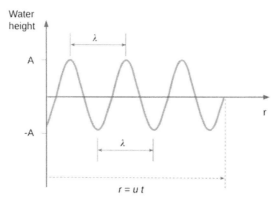

Figure 8.6: Water profile at a specific time t from the moment the stone fell inside the water. The oscillation of frequency v at the origin, $r = 0$, generates a wave of wavelength λ. The wave reaches a maximum height A and a dip of minimum height $-A$.

We know that the edge of the wave has propagated a distance $r = ut$ during the time t, where u is the phase velocity. At the edge of the wave, the height is zero. Then, we can write the following

$$h(ut) = 0 = A \cdot \sin\left(\frac{2\pi ut}{\lambda} + \phi(t)\right) \tag{8.4}$$

The condition above is valid when the term in the parentheses equals zero,

$$\phi(t) = -\frac{2\pi ut}{\lambda} \tag{8.5}$$

As we discussed earlier the phase velocity is $u = \lambda v$, then

$$\phi(t) = -2\pi vt \tag{8.6}$$

Using this expression for $\phi(t)$, the wave equation can be written as

$$h(r,t) = A \cdot \sin\left(\frac{2\pi r}{\lambda} - 2\pi vt\right) \tag{8.7}$$

In comparison to a simple harmonic oscillator, equation 8.7 is a function of space and time. It shows how the wave amplitude (height of the water surface) develops in time across the whole space. In other words, it represents how the wave propagates.

Oscillator as a wave generator

At the origin of our coordinate where the stone fell in water, r = 0, the wave equation is reduced to a simple harmonic oscillation.

$$h(0, t) = A \cdot \sin(2\pi v t) \tag{8.8}$$

These oscillations at the origin generated a wave that propagates away from the center along the radial direction due to water's surface tension. We can then think about it as follows: when the stone fell in the water it formed an oscillator. The oscillator became a source that generates mechanical waves. By mechanical waves we mean a pattern of repeated changes in the mechanical properties of a medium which changes in time and space. In our example, the mechanical properties of the water surface level are pushed upwards and pulled downwards in a sinusoidal pattern. The wave propagated in a medium, which is water in our case. We could say that we need a medium for the wave to propagate, or if there was no water there would be no wave. This is correct. However, we will show later that there exists another type of waves that do not require a medium, which are electromagnetic waves. As a matter of fact, these waves caused a large debate between scientists to the point that a medium called "ether" was invented to solve this problem. This was later proven to be wrong and electromagnetic waves as of today are assumed to propagate without medium such as almost the case in space, at least in classical physics. We will introduce electromagnetic waves in the next unit.

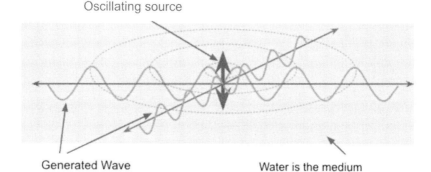

Figure 8.7: Mechanical waves that are generated by an oscillating source and propagate in a medium.

Back to our mechanical wave, we can say the following. There is a source which is an oscillator. The source generates waves that are formed by a repeated pattern of change in the mechanical properties (water level here) in a medium as illustrated in figure 8.7. We can think of another example of mechanical waves and wave propagation, sound. When we talk, our mouth becomes an oscillator. It presses and relaxes the air molecules in some periodical pattern. This pressing and relaxation propagates in air and reaches another person's ears. That is then transformed into electrical signals in our hearing system which is then translated into sound by our brain. These mechanical waves formed in air are commonly known as sound waves or acoustic waves. Notice that sound waves as well travel in solids and liquid.

A third example of a mechanical wave is earthquake. The sliding of the earth tectonic plates creates an oscillator, and this oscillator generates mechanical waves in the earth surface. These waves propagate away from the center of the quake the same way the water waves propagate from the stone. These waves would travel large distances, and we can feel the shaking of the ground due to the nature of wave propagation. These waves are commonly called Seismic waves. Table 2.1 summarizes these examples of some mechanical waves and their propagation.

Table 8.1: Examples of mechanical waves.

Wave name	Type	Source	Medium
Water waves	Mechanical	Oscillations due to objects falling, wind or quakes	Water
Sound waves/acoustic waves	Mechanical	Oscillations due to pressing gas, solid or liquid.	Gas, solid or liquid
Seismic waves	Mechanical	Oscillations due to sliding of earth's tectonic plates	Earth surface

Long distance of propagation

Looking at the wave generated by a stone in figure 8.5, a question arises: "what does the wave look like when it reaches a faraway shore?" To answer this, we need to follow the wave as it propagates further and further away from the origin. It is not hard to use a computer and plot equation 8.7 over space at a specific instant. These calculations are visualized in figure 8.8. When the wave propagates far away from the origin and when observing it around a small region, the curvature of the lines (bright lines representing peaks or dark lines representing dips) tends to disappear resulting in almost straight lines. That could explain why at the shore, the sea waves appear to be on straight lines as the waves have traveled far distance from the origin of oscillation that generated them.

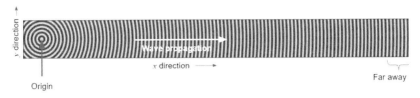

Figure 8.8: Wave propagation for a long distance away from the origin.

When the wave reaches the shore, as many of us have observed on many occasions, it exerts force on the objects at the shore such as rocks, beach sand or any other objects. Hence one could say that the wave has transferred the effect of the source (stone falling in water) to an object that is far away thanks to its propagation. Simply, the wave transfers the energy from the source to a remote receiver.

This is exactly what happens when we call on a person standing far. Our mouth becomes an oscillator that generates acoustic waves. The waves travel to the receiver's ears. The acoustic waves exert force on the eardrums. The vibrations in the eardrum are transferred to electrical signals which the brain translates to sound as illustrated in figure 8.9.

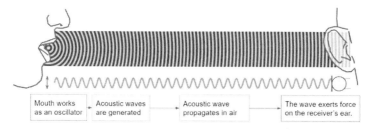

| Mouth works as an oscillator | Acoustic waves are generated | Acoustic wave propagates in air | The wave exerts force on the receiver's ear. |

Figure 8.9: Acoustic waves produced by the speaker's mouth propagate to the receiver's ear and exert force on the eardrum. That causes electrical signals to be generated and translated by the brain as sound.

Example 8.1:

We know that the speed of sound in air is $u = 343$ m/s and that human can hear sounds with frequencies of oscillations between 20 $Hz < v < 20,000$ Hz. What is the wavelength range of human hearing?

We know the relation between wavelength and frequency

$$u = \lambda v \rightarrow \lambda = v/u$$

So, for the minimum frequency, the wavelength is

$$\lambda = 343/20 = 17.5m$$

At the maximum frequency, the wavelength is $\lambda = 343/20,000 = 17.5mm$

Then the wavelength of human hearing capability is $17.5mm < \lambda < 17.5m$

Angular frequency and wavenumber

Equation 8.7 can be written in a more compact form as

$$h(r,t) = A \cdot \sin(kr - \omega t) \qquad (8.9)$$

Where, $k = 2\pi/\lambda$, is known as the wavenumber. It has units of m^{-1}. The term $\omega = 2\pi v$ is known as the angular frequency and has units of radian/s. So, what is the benefit of these two terms? One could say, if the wave propagates for a distance r, then the phase gained by that is obtained by simply multiplying the distance by the wavenumber. Similarly, if the source oscillates for a time t, that causes a phase change that equals the multiplication of time by the angular frequency. Let us now examine the relationship between the two quantities. We know that relation between wavelength and frequency as $\lambda = u/v$. If we multiply and divide the right side by 2π we obtain

$$u = \lambda v = \frac{2\pi v}{2\pi/\lambda} = \frac{\omega}{k} \qquad (8.10)$$

The phase velocity of the wave is the division of the angular frequency by the wavenumber. One interesting observation to mention here is that the wavenumber has units of 1/m. That can be compared to the unit of frequency 1/s. That is why we commonly refer to the wavenumber as the amplitude of the spatial frequency. Here, we call it spatial to refer to frequency in space compared to the term frequency alone which commonly refers to time oscillations.

Example 8.2:

What is the wavenumber and angular frequency for a sound wave that oscillates at 1000 Hz.

The sound phase velocity is $u = 343$ m/s. Given the frequency is $v = 1000$ Hz

The angular frequency is $\omega = 2\pi v = 2\pi \times 1000 = 6283$ rad/s

The wave number can be obtained from $u = \omega/k \rightarrow k = \omega/u = 6283/343 = 18.31 m^{-1}$

8C. More on waves

Figure 8.8 shows a top view of a wave propagation on the surface of the water after a stone was dropped at the origin. The dark lines represent the dip in the water surface while the bright ones represent the peaks.

Types of waves

Near the origin the dip and peak lines form circles centered around the origin. If we pick one of these circles, we will observe that the water level is the same at all points on it. If we fix the circle location and keep monitoring the water level as time progresses then we would observe that for all the points on the circle, the surface is moving vertically up and down along the z-direction in the exact same way. For this reason, this wave is commonly referred to as a circular wave. This is because for any circle centered around the origin all the points on that circle act in the same way. The shape over which all the points act in the same way is commonly known as the wave front. For example, the height of the water surface is the same at all the points on a **circle** centered around the origin. Hence, the water surface wave has a circular wave front. The shape of this wave front defines the type of the wave. Circular wavefronts define circular waves. But are there any other wave fronts?

Let us examine the acoustic wave in figure 8.9. Even though we draw the acoustic waves in a two-dimensional space, we know that these waves propagate in volume. A person sitting next to the speaker and even behind him will still be able to hear him talking. So, in this case we can approximate the mouth oscillations as an oscillator located at one point (the origin) and generate waves that propagate in the radial direction in the three-dimensional space as illustrated in figure 8.10. The wavefronts in this case are concentric spheres centered around the origin of oscillations. Hence, the generated waves are referred to as spherical waves. Mathematically we can write the strength of the pressure at any point in the volume at any instant of time as

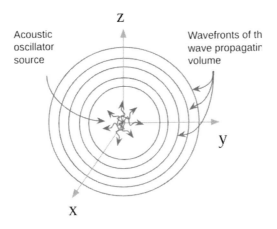

Figure 8.10: An acoustic wave generated in the volume by an oscillating source placed at the origin.

$$P(r,t) = A \cdot \sin(kr - 2\pi vt) \tag{8.11}$$

where in this case $r = \sqrt{x^2 + y^2 + z^2}$. The constant A here represents the maximum value of pressure applied by the source on the surrounding gases. Let us now place a microphone at a location along the x axis that is far away from the source. The microphone receives acoustic waves across its area, a, which extends in the y-z plane. If the dimensions of this area are much smaller than distance x, then one neglects the contribution of both y and z values on the amplitude of the displacement vector and $r \approx x$.

Example 8.3:

A microphone has a reception area defined by width × height of 2mm × 2mm is placed on the x-axis at a distance of 2 away from the audio source that oscillates at a frequency of 1000 Hz. If we constrain ourselves to the x-y plane (place z=0) can you prove that for the given dimensions the y effect can be neglected in the wave?

$x = 2m$ and $y_{max} = 2 \times 10^{-3} m$

Speed of sound $u = 343$ m/sec; This is a known fact

$v = 1000 Hz \rightarrow \lambda = u/v = 343/1000 = 0.343m$

Wavenumber, $k = 2\pi/\lambda = 2\pi/0.343 = 18.32 m^{-1}$

At the center, when y =0 and assuming t=0

$r = x = 2m$

$P = A \cdot \cos(kr) = A \cdot \cos(18.32 \times 2) = 0.86218389728 \times A$

At the maximum height of the microphone receiver

$r = \sqrt{2^2 + 0.002^2} = 2.000001 m$

$P = A \cdot \cos(kr) = A \cdot \cos(18.32 \times 2.000001) = 0.86218853723 \times A$

As seen in example 8.3, the pressure value at the center and the edge of the microphone receiver is almost identical (till the fifth decimal location.) Hence, we can assume that the wavefront across the microphone is a plane normal to the x axis (y-z plane). Hence, having a planar wavefront defines the wave to be a planar wave. In this example we can write the planar wave equation as

$$P(r,t) = A \cdot \cos(kx - 2\pi vt) \tag{8.12}$$

Notice here, the value of the pressure at any point inside the y-z plane (normal to the x-direction) is the same as the expression in 8.12, which is independent of y and z. That gives an infinite extension of the wavefront along the y and z directions. This of course is not a realistic assumption. Hence, these waves do not exist in nature, but they are rather an approximation to finite wavefronts far away from the source of the oscillations. In spite of this fact, plane waves provide a great simplification when solving most of the wave propagation problems in physics. Later, when the students become more advanced in mathematics, they will learn that any wavefront could be expanded into an infinite number of plane waves. This is commonly known as Fourier expansion. This is however outside the scope of this book. Instead, we will examine the properties of plane waves and show how the wave direction can be introduced inside the expression of the wave function.

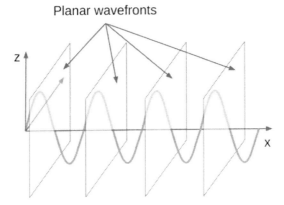

Figure 8.11: Planar wave propagating in the x direction with planar wavefronts.

Plane waves

A plane wave propagating along the x axis is represented by equation 8.12 where the wave amplitude is changing in a periodical pattern along that direction. The wavefronts form planes normal to the x-axis as depicted in figure 8.11.

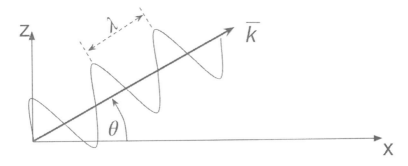

Figure 8.12: Planar wave propagating in the x-z plane with an angle θ with respect to the x-axis.

What if the plane wave is propagating along a direction θ in the x-z plane as depicted in figure 8.12? How can we mathematically express this wave? To answer this, we are going to use some trigonometrical trick. Equation

8.11 describes the wave amplitude along the direction of the wave propagation. Hence, we can use the same expression here. Though we will apply the following trick

$$P(r,t) = A \cdot \cos(kr \cdot (\cos^2\theta + \sin^2\theta) - 2\pi vt) \qquad (8.13)$$

This is the same equation as 8.11 knowing that $cos^2\theta + sin^2\theta = 1$. We can rearrange the expression above as follows

$$P(r,t) = A \cdot \cos\big((r\cos\theta \cdot k\cos\theta + r\sin\theta \cdot k\sin\theta) - 2\pi vt\big) \ (8.14)$$

We know from the geometry of the problem that the displacement vector in our case is $\bar{r} = (x, y, z)$. When the propagation is in a direction inside the x-z plane with an angle θ relative to the x-axis then we can write that $x = r\cos\theta$, $y = 0$ and $z = r\sin\theta$. Similarly, we would assume a vector $\bar{k} = (k_x, k_y, k_z)$ that we commonly refer to it as the wave vector. The wave vector has an amplitude of the wavenumber and direction along the direction of the wave propagation. Hence, in our example we can write the x and z components of the vector as

$$k_x = k\cos\theta \qquad (8.15a)$$

$$k_y = 0 \qquad (8.15b)$$

$$k_z = k\sin\theta \qquad (8.15c)$$

We can then write equation 2.14 as

$$P(r,t) = A \cdot \cos\big(k_x x + k_y y - 2\pi vt\big) \qquad (8.16)$$

This equation can be written in terms of a dot product between the displacement vector and the wavenumber as

$$P(r,t) = A \cdot \cos\big(\bar{k} \cdot \bar{r} - 2\pi vt\big) \qquad (8.17)$$

Hence, the representation of the plane wave contains its direction in terms of the wave vector as well as the frequency/wavelength information. Remember that the wavenumber is the amplitude of the wave vector or

$$k = |\bar{k}| = \sqrt{k_x^2 + k_y^2 + kz^2} \qquad (8.18)$$

Example 8.4

Write an expression for an acoustic wave that propagates far away from the source where the receiver is in the x-y plane at an angle 30° to the x axis. The wave has a frequency of 2000 Hz.

Type of wave is a plane wave, as it propagates far away from the source

Direction: $\theta = 30^o$ from the x-axis. Frequency: $v = 2000\ Hz$

Speed of sound: $u = 343\ m/s$

The wavelength is: $\lambda = u/v = 343/2000 = 0.1715\ m$

The wavenumber: $k = 2\pi/\lambda = 36.64\ m^{-1}$

The wave vector:

$k_x = k\cos\theta = 36.64\cos(30^o) = 31.73\ m^{-1}$
$k_y = k\sin\theta = 36.64\sin(30^o) = 18.32\ m^{-1}$
$k_z = 0$

Wave expression:

$$P(r,t) = A \cdot \cos(31.73x + 18.32y - 2\pi \times 2000t)$$
$$= A \cdot \cos(31.73x + 18.32y - 12566.3t)$$

Example 8.5:

An operator recorded a seismic wave on his computer as the center of the earthquake was far, he found the following plane wave expression:

$\cos(0.019x + 0.0089y - 125.664t)$

Can you find information regarding the frequency of oscillation, speed of the wave and direction of propagation?

Wave vector: $k_x = 0.019m^{-1}$, $k_y = 0.0089m^{-1}$, $k_z = 0m^{-1}$

Angular frequency: $\omega = 2\pi v = 125.664 rad/s$

Wavenumber: $k = |k| = \sqrt{k_x^2 + k_y^2 + K_z^2} = \sqrt{0.019^2 + 0.0089^2 + 0^2} = 0.02117 m^{-1}$

Wavelength: $k = 2\pi/\lambda \rightarrow \lambda = 2\pi/k = 2\pi/0.02117 = 296.77 m$

Frequency: $v = \omega/2\pi = 125.664/2\pi \approx 20 Hz$

Wave speed/Phase velocity: $u = \lambda v = 296.77 \times 20 = 5935.4 m/s$

Direction:

$k_x = k \cos\theta$ and $k_y = k \sin\theta$, dividing both $\dfrac{k_y}{k_x} = \dfrac{\sin\theta}{\cos\theta} = \tan\theta$

$$\theta = \tan^{-1}(k_y/k_x) = \tan^{-1}(0.0089/0.019) = 25.1^o$$

8D. Summary

- Wave is a repeated pattern in space that changes in time.
  - It is generated by an oscillator of frequency v (Hz)
  - It propagates in space with speed u (m/s)
  - The speed at which the wave propagates is called phase velocity
  - The period in space is the wavelength λ (m)
    - $\lambda v = u$
- Mathematically, we can write a wave propagating out of a point source
  - $h(r, t) = A \cdot \sin\left(\frac{2\pi r}{\lambda} - 2\pi v t\right)$
  - A is the strength of the wave
- At the origin, $r = 0$, the oscillator generates the wave
  - $h(r, 0) = A \cdot \sin(-2\pi v t)$
- We can write the wave as well as
  - $h(r, t) = A \cdot \sin(kr - \omega t)$
  - $\omega = 2\pi v$ is the angular frequency as it has units of *rad/s*
  - $k = \frac{2\pi}{\lambda}$ is the wavenumber and it has units of m^{-1}
- Many of the waves we observe in our daily life are mechanical waves.
  - Mechanical waves require a medium to propagate
    - Water surface waves propagate in water
    - Sound waves propagates in air (full of gas molecules)
    - Seismic waves propagate on the surface of the earth.
- There are types of waves which do not require a medium to propagate such as electromagnetic waves and light.
- Near the oscillator source, the wave is called circular (surface wave) or spherical (volume wave).

- Far from the source, we can approximate the wave as
    - $P(r,t) = A \cdot \cos(\overline{k} \cdot \overline{r} - 2\pi\nu t)$
    - This is referred to as plane wave
    - $\overline{k}$ is the wave vector
        - $\overline{k} = \left(k_x,\ k_y, k_z\right)$
        - The direction of $\overline{k}$ is
            - the direction of the plane wave propagation
        - The amplitude of $\overline{k}$ is
            - $\left|\overline{k}\right| = \sqrt{k_x^2 + k_y^2 + k_z^2} = \frac{2\pi}{\lambda}$

8E. Review questions

1 – Complete the table

Wavelength (mm)	Frequency (Hz)	Speed (m/sec)
100	6000	
	30	334
2000		5000
0.1	3109	

2 – Calculate the speed of the following waves.

   a) $10 \cdot \cos(6283\,t - 18.9\,x)$
   b) $3 \cdot \cos(94247.7\,t - 0.418\,x)$
   c) $0.5 \cdot \cos(314\,t - 0.628\,x)$

3 – If an electrical buzzer produces an audio frequency of 10000 Hz. Calculate the wavelength of the produced wave when it propagates in:

   a)   Air   b) Water   c) Wood

Note: Look up the speed of sound in each of the mentioned media.

4 – Which of the following wave equations represents sound waves propagating in air? Express the reason.

   a) $0.5 \cdot \cos(314\,t - 9.15\,x)$
   b) $10 \cdot \cos(314\,t - 0.915\,x)$
   c) $4 \cdot \cos(314\,t - 0.0915\,x)$

5 – The plots below show a wave at two different time instances separated by a time step of 400 msec. Estimate the wavelength, speed and frequency of the wave then write a mathematical expression for the wave.

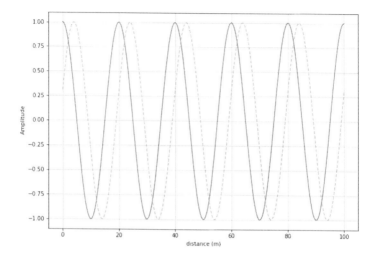

6 – An LC circuit is used as an oscillator that generates waves (Electromagnetic waves), if the wave propagates at the speed of light write an expression for the wave when the following components are used

C (F)	L (mH)	Frequency (Hz)	Wavelength (m)	Expression
0.2	0.3			
0.04	0.2			
0.01	0.01			

CHAPTER NINE

BASICS OF ELECTROMAGNETICS

9A. Electrical oscillators and waves

In the chapter of oscillators, we introduced a non-mechanical oscillator that consists of a capacitor and a coil. The initially stored energy in the capacitor (static charges on the plates) is transferred to the coil (moving charges in the wire) and back to the capacitor in a repetitive way. The generated a voltage drop across the capacitor that has a sinusoidal form as

$$V(t) = V_o \cdot \cos\left(\frac{t}{\sqrt{LC}}\right) \tag{9.1}$$

That generates an electrical oscillator where charges move between the two components in a periodical pattern. Similar oscillating response can be obtained when placing the two components in series.

Series LC circuit

The parallel LC circuit in the previous chapter can be as well modified without the loss of the concept in a series form as in figure 9.1 in comparison to the parallel circuit, explained earlier.

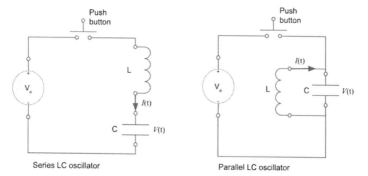

Figure 9.1: Series LC oscillator circuit compared to parallel LC circuit.

We know that in the parallel circuit if we press the push button for a time that is long enough then the capacitor can be charged such that a potential difference of V_o is built across it. Now, if we release the push button, the energy will be moving back and forth between the capacitor and the inductor in an oscillatory form with an angular frequency of $\omega = 1/\sqrt{LC}$. The oscillation sustains the condition that the total voltage drop across the capacitor and inductor is zero, or $V_C + V_L = 0$. This can be as well the case for the series circuit when we press the button for a time that is long enough to charge the capacitor then release it. Again, the total voltage of the capacitor and the inductor should equal zero as there is no electrical potential applied across the circuit when the button is released. In this case an oscillator is formed, and we obtain the same oscillation angular frequency of $\omega = 1/\sqrt{LC}$. Notice that here we neglected the damping effect due to resistance for the purpose of simplicity.

In previous chapter we found that oscillators work as sources which generate waves. Like the case of a stone in the water, the oscillating water level at the center generates waves that propagate along the water surface via the surface tension. In all the mechanical waves examples introduced, there has always existed a medium through which the waves propagate. The questions here are, what type of waves that the electrical oscillator could possibly generate? and what medium would be used to carry these waves?

Electrical oscillator and waves

Like the water waves, one can think that the electrical oscillator generates some sort of waves as well. These waves propagate somehow and reach some remote charges. The charges could be originally stationary or moving. The waves then exert a force on the charges similar to the effect the water waves induce on the remote beach sands and rocks as visualized in figure 8.2. So, what is the nature of these waves and how do they propagate?

To answer this question, one needs to find a law that governs the force between two remote charges. Thankfully, such a law exists, and it was driven experimentally in 1785 by a French physicist. Many students might know it by heart, that is Coulomb's law of electrostatics.

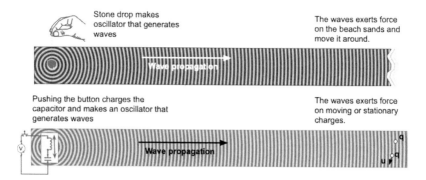

Figure 9.2: Similar to water waves that propagate and exerts force on the beach sand, one can think of the electrical oscillator to generate waves that exert force on remote charges.

Coulomb's law

It is a law that is named after a French physicist Charles-Augustin de Coulomb who published it in 1785. Similar to Newton's gravitational force between two masses, a force of attraction or repelling between two charges was observed experimentally to be inversely proportional to the square of the distance between the two charges. Mathematically, Coulomb formulated the following expression for the force

$$\bar{F} = k_e \frac{q_1 q_2}{r^2} \hat{r} \tag{9.2}$$

Here, q_1 and q_2 are the magnitudes of the two charges with the proper sign (− for negative charge and + for positive charge). Charge magnitude has a unit of Coulomb or C in short. Originally 1 C was the total charge received by an electrical current of 1 Amp in 1 second. One electron (or one proton) carries an elementary charge amplitude of

$$q_e = 1.602176634 \times 10^{-19} C$$

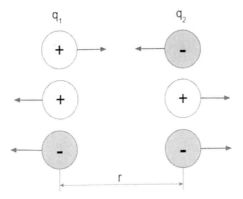

Figure 9.3: The force between two different charges depends on the signs of the charges and their separation. Opposite charges attract each other while charges having the same sign rebel away from each other.

Electron charge is negative while proton's charge is positive. For the force in equation 9.2 the constant k_e is the Coulomb's constant and it has a magnitude of

$$k_e \approx 8.988 \times 10^9 N \cdot m^2 \cdot C^{-2} \tag{9.3}$$

The unit vector $\hat{r}$ points from one charge to the other. So, if we are looking at the force on q_2 due to q_1 then the unit vector is pointing from q_1 towards q_2. If the two charges have the same sign (positive and positive or negative and negative) then the total force is positive and q_2 will experience a force that pushes it away from q_1. If the two charges have opposite signs, then the force is negative and q_2 will experience a pull towards q_1 as visualized in figure 9.3.

Rise of the electric field

One might bonder on this effect for a while and ask the following question, how can a charge exert a force on another that is remotely placed. What did carry this force? To make a model that gives a one-on-one sort of interaction, we would need to bring back a quantity we called the electric field in chapter seven and give it a rather proper introduction. In there, we have mentioned that the electric field is the mean by which static charges are stored on the capacitor plates. We did not however discuss how this quantity was introduced and how it keeps the charges on the capacitor plates. To make the connection, let us start by rearranging equation 9.1 when grouping all the terms except for q_2.

$$\overline{F} = q_2 \left(\frac{k_e q_1}{r^2}\right) \hat{r} \qquad (9.4)$$

Equation 9.4 now reads as follows: the force that exerts on charge q_2 due to the presence of q_1 is the multiplication of the charge amplitude and some vector quantity.

$$\overline{E} = \left(\frac{k_e q_1}{r^2}\right) \hat{r} \qquad (9.5)$$

This quantity is what we will call the electric field. We can hence say that the charge q_1 produced an electric field $\overline{E}$. The electric field travels to q_2 and then exerts a force $\overline{F}$ on it.

$$\overline{F} = q_2 \overline{E} \qquad (9.6)$$

So, we can state the following scenario:

- At time t, the charge q_1 emits an electric field.
- At time $t + \tau$, the electric field reaches the charge q_2 with an amplitude that is reduced by the square of the distance ($|\overline{E}| = k_e q_1 / r^2$).
- The electric field exerts a force on q_2 ($\overline{F} = q_2 \overline{E}$).

The time $\tau = r/u_p$ where u_p is the speed of the electric field propagation. This model is illustrated in figure 9.4. Based on special relativity, the speed of light is the maximum speed that can be reached and serves as a ruler in the universe. Hence one cannot expect that the electric field to propagate faster than the speed of light. Even though Coulomb's law deals with electrostatic charges and we might get the sense that the effect is instantaneous, however the delay τ between the field emission and the force to be exerted is essential to keep the statement that no object can travel faster than the speed of light valid. We could then safely assume that the electric field travels at the maximum valid speed, which is the speed of light or $u_p = c = 299792458\ m/s$.

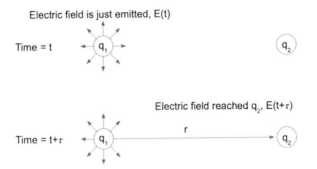

Figure 9.4: Charge q_1 emits an electric field at time t and the field reaches charge q_2 at time $t + \tau$ where c is the speed of light at which the electric field propagates.

Before proceeding further, one needs to revisit equation 9.6. We could arrange it differently defining the electric field as the force divided by the charge, $\overline{E} = \overline{F}/q_2$. In this case the electric field clearly has units of N/C. We know that work equals the amount of energy needed to move an object over a distance against an applied force. Hence, the unit of force equals that of energy divided by distance, $N = J/m$, where J stands for Joule, the unit of energy. The unit of the electric field is then $N/C = J/(m \cdot C)$. We know from the definition of the electromotive force, which has units of Volts (or V), that it is the amount of work done on a unit charge. Hence, $V = J/C$. Using this, one can easily deduce that the units of the electric field can be as well V/m. This agrees well with the definition of the electric field across the capacitor, $E = V/d$, where d is the capacitor separation distance and V is the potential difference across the capacitor plates.

> The electric field has units of N/C = J/m C = V/m

The model in figure 9.4 looks in a way similar to throwing a stone in water, the oscillating center generates waves, and the waves exert force on remote beach objects. However, the electric field in equation 9.5 does not resemble by any form a wave. This is clear from the fact that Coulomb's law considers static charges with no oscillations. It gives us an expression for the force exerted by one fixed charge on another fixed charge that is remotely located. This limit is typically referred to as **electrostatics**. So, in electrostatics, we deal with fixed charges. However, in the case of the stone in water example we talked about an oscillating source that generates waves. What happens then if charge q_1 starts to oscillate?

Oscillating charge

If we assume that the charge q_1 starts to oscillate up and down, along the y axis for instance, then at any moment of time, the charge will have a velocity vector $u(t)$ and an acceleration vector $a(t)$. Both vectors are along the y-axis. To find a mathematical representation of the electric field of the moving charge in terms of velocity and acceleration we could start with a model similar to that in figure 9.4. However, there is a slight difference we need to account for, that is the fact when q_2 receives the electric field, q_1 has already moved away from its original location.

- At time t, the charge q_1 emits an electric field $E(t)$.

- At time $t + \tau$, the electric field reaches q_2 and exerts a force on it.

$$\overline{E}(t + \tau) = \frac{k_e q_1}{r(t)^2} \hat{r}(t) \tag{9.7}$$

Notice that we use the unit vector $\hat{r}(t)$ in equation 9.7 not $\hat{r}(t + \tau)$. That is because the field that reaches q_2 is the one generated by q_1 at time t as illustrated in figure 9.5. As shown in the figure, during the delay τ the charge q_1 has moved by a displacement of $\overline{\delta r}$ as it oscillates.

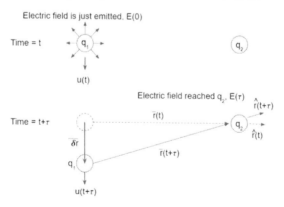

Figure 9.5: Electric field emitted by q_1 at time t reaches q_2 at time $t + \tau$. During the propagation time, the charge q_1 moves by a displacement of $\overline{\delta r}$.

The unit vectors can be written as

$$\hat{r}(t) = \overline{r}(t)/r(t) \tag{9.8a}$$

$$\hat{r}(t + \tau) = \overline{r}(t + \tau)/r(t + \tau) \qquad (9.8b)$$

Where $r(t) = |\overline{r}(t)|$ and $r(t + \tau) = |\overline{r}(t + \tau)|$. It is however more adequate to represent the force in terms of the displacement vector at the same time instance, $\overline{r}(t + \tau)$. To do that, we would need to use a known trick of Taylor expansion. We have introduced Taylor expansion in an earlier chapter. Here, we will expand the displacement vector $\overline{r}(t + \tau)$ as follows

$$\overline{r}(t + \tau) = \overline{r}(t) + \tau \cdot \frac{d\overline{r}(t)}{dt} + \frac{\tau^2}{2} \cdot \frac{d^2\overline{r}(t)}{dt^2} + \dots \qquad (9.9a)$$

If we assume that the distance $r(t)$ is large enough such that the delay τ remains very small, then we could neglect the terms with power higher than τ^2 in equation 9.9a. Neglecting higher order derivatives of r, we can write the following expression

$$\overline{r}(t) \approx \overline{r}(t + \tau) - \tau \cdot \overline{u}(t) - \frac{1}{2}\tau^2 \cdot \overline{a}(t) \qquad (9.9b)$$

where $\overline{u}(t) = d\overline{r}(t)/dt$, which is the velocity of the charge at time t. The vector $\overline{a}(t) = d^2\overline{r}(t)/dt^2$ and that is the acceleration of the charge at time t. We have discussed earlier that the electric field travels at the speed of light and hence the delay is $\tau = r(t)/c$.

$$\overline{r}(t) \approx \overline{r}(t + \tau) - \frac{r(t)}{c} \cdot \overline{u}(t) - \frac{r(t)^2}{2c^2} \cdot \overline{a}(t) \qquad (9.9c)$$

Let us assume that the separation between the two charges is large such that the change in the amplitude is negligible. We could then crudely state that $r(t) \approx r(t + \tau) = r$. The unit vectors can be then written as $\hat{r}(t) = \overline{r}(t)/r$ and $\hat{r}(t + \tau) = \overline{r}(t + \tau)/r$. Using these assumptions and substituting equation 9.9c into 9.7 we obtain the following

$$\overline{E}(t + \tau) = \frac{k_e q_1}{r^2}\hat{r}(t) = \frac{k_e q_1}{r^3}\overline{r}(t) \qquad (9.10a)$$

$$\overline{E}(t + \tau) = \frac{k_e q_1}{r^3}\overline{r}(t + \tau) - \frac{k_e q_1}{r^2 c}\overline{u}(t) - \frac{k_e q_1}{2rc^2}\overline{a}(t) \qquad (9.10b)$$

Notice that the first term is the original Coulomb law between the two charges at the new position when no motion is considered. The other two terms however depend on the particle's motion. Comparing the three terms in equation 9.10, we notice that the last term reduces with a factor of $1/r$ while the second is reduced by $1/r^2$ and the first is reduced by a factor $1/r^3$. If we move the charge q_2 far away from the oscillating source q_1 then we expect that after some distance the first two terms will die out and the last term becomes the most significant. We could hence safely state that in the far field we can drop the first two terms in equation 9.10b and approximate the electric field as

$$\overline{E}(t + \tau) \approx -\frac{k_e q_1}{rc^2}\overline{a}(t) \tag{9.11}$$

One needs to pay attention that in reaching the expression in 9.11, we have applied a crude approximation that the distance r is very large such that the charge motion during the time τ has an effect mainly on the direction of the field with minimal modification in the field's amplitude. In other words, we assume that $|r(t + \tau)| \approx |r(t)| = r$. This assumption is commonly referred to as the **far field approximation**. It is important to notice that at the far field the electric field is along the direction of oscillation of the charge and its amplitude depends on the acceleration of the charge. This is an interesting result which has the following indications:

1.   Having an oscillating (moving) charge generates an electric field that propagates for a longer distance compared to stationary charges. This is as the amplitude reduces by a factor of $1/r$ compared to $1/r^3$ in the case of electrostatics.

2.   The direction of the electric field is along the direction of the charge oscillation, not along the displacement vector between the two charges.

If the charge is oscillating in a sinusoidal form along the y direction, then we can say it has displacement vector, velocity and acceleration as

$$\overline{r}(t) = r_o \cos(\omega t)\,\hat{y} \tag{9.12a}$$

$$\overline{u}(t) = \frac{d\overline{r}(t)}{dt} = -\omega r_o \sin(\omega t)\,\hat{y} \tag{9.12b}$$

$$\overline{a}(t) = \frac{d^2\overline{r}(t)}{dt^2} = -\omega^2 r_o \cos(\omega t)\,\hat{y} \tag{9.12c}$$

Here, r_o is the maximum displacement the charge could move during the oscillation. Substituting equation 9.12c into equation 9.11 we obtain the following field expression

$$\overline{E}(t + \tau) = \frac{k_e q r_o \omega^2}{rc^2} \cos(\omega t)\,\hat{y} \tag{9.13}$$

Finally, it would be more adequate to present the electric field at time t instead of $t + \tau$. To do so, we need replace t by $t - \tau$ in equation 9.13.

$$\overline{E}(t - \tau + \tau) = \overline{E}(t) = \frac{k_e q r_0 \omega^2}{rc^2} \cos(\omega(t - \tau)) \qquad (9.14)$$

We know that $\tau = r/c$. To simplify the expression in 9.14, we could define the following constant potential constant $A = \frac{k_e q r_0 \omega^2}{2c^2}$. Using the expression of the delay and using the new constant A, equation 9.14 can be written as

$$\overline{E}(t) \approx \frac{A}{r} \cos\left(\omega t - \frac{\omega}{c} r\right) \hat{y} \qquad (9.15)$$

In chapter eight, we introduced the wavenumber k that equals the angular frequency divided by the phase velocity. Here, the phase velocity or the speed of the wave propagation equals the speed of light. Hence, for the electric field the propagation constant is $k = \omega/c$. We can now write the electric field to include time and space dependency as

$$\overline{E}(t) \approx \frac{A}{r} \cos(\omega t - kr) \hat{y} \qquad (9.16)$$

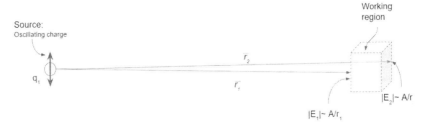

Figure 9.6: Far field working region where the change of the displacement vector has minimal effect on the field amplitude.

Let us assume that the field is far enough from the source such that the changes of r across the working region has minimal effect on the field's amplitude as visualized in figure 9.6. The amplitude of the electric field closest to the source is proportional to $1/r_1$ while at the furthest surface it is proportional to $1/r_2$. If the distance $r_2 = r_1 + \delta$ where $\delta \ll r_1$ then we can safely assume that $\frac{1}{r_2} = \frac{1}{r_1 + \delta} \approx \frac{1}{r_1}$. To examine this assumption, let us assume that $r_2 = 1.001 \times r_1$. Hence, $\frac{1}{r_2} = \frac{1}{1.001 \times r_1} = \frac{0.999}{r_1} \approx \frac{1}{r_1}$. Based on this discussion, we can assume that the term A/r remains constant across the proposed working region, and we can replace it with a constant $E_o = A/r$ or

$$\overline{E}(t) = E_o \cdot \cos(\omega t - kr)\,\hat{y} \tag{9.17}$$

This expression is very close to the one we obtained by a stone dropping in a water case. The electric field produced by an oscillating charge is in the form of a wave that has a frequency $v = \omega/2\pi$. From the previous chapter we know that the wavelength multiplied by the frequency equals the phase velocity. Hence, for this specific wave the wavelength frequency multiplication is

$$\lambda v = c \tag{9.18}$$

This type of wave, produced by oscillating charges and travels at the speed of light is commonly known as **electromagnetic waves** and the generation of these waves is commonly referred to as **electromagnetic radiation**.

9B. Electromagnetic waves

During the discussion in the previous section and in Coulomb's law formulation we did not talk about a presence of a medium between charges q_1 and q_2. Unlike the mechanical waves, there was no mentioning of a mechanism by which these waves start to propagate away from the center of oscillations. This was a topic of debate in the scientific society during the 19th century. That resulted in an introduction of a medium called "the ether" which was assumed to fill in the space surrounding the earth and carry the electromagnetic waves in a way similar to water carrying surface waves. This concept however was discarded by the beginning of the twentieth century and electromagnetic waves since then are assumed not to require a medium for propagation. They do propagate in vacuum.

Electromagnetic waves properties

Electromagnetic waves propagate in vacuum and travel with the speed of light. The properties of the electromagnetic waves till now can be summarized in table 9.1.

Table 9.1: Basic properties of electromagnetic waves

Property	Description
Oscillating source	Acceleration of the electric charges
Oscillating nature	Amplitude of the electric field
Medium to transfer	No medium needed
Phase velocity	Speed of light, c=299792458 m / s
Wavelength, frequency relation	$\lambda v = c$

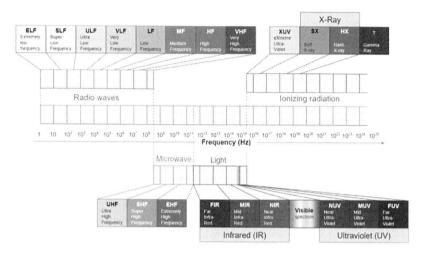

Figure 9.7: Electromagnetic spectrum showing the different classes of radiation. Here, the frequency is presented on a logarithmic scale.

By definition, the frequency of the electromagnetic waves can be from few Hz up to more than 10^{25} Hz. Each band of frequencies defines a specific class of radiation. The division of the classes over the frequency range is commonly known as the electromagnetic spectrum as illustrated in figure 9.7. Notice that in the figure frequency is plotted in a log scale.

Radio waves cover a large range of frequencies from 3 Hz to 300 MHz. Here "M" stands for mega or 10^6. As the name indicates, radio waves are commonly used for wireless communication and signal broadcasting such as radio and TV. In terms of wavelength, the radio waves can be from 108 m down to 1 meter.

Moving to higher frequencies, microwaves cover a range from 300 MHz to 300 GHz. Here "G" stands for giga or 10^9. That gives wavelengths from 1 m down to 1 mm. Increasing the frequency of radiation we reach the light spectrum. Light is divided into three main regions: Infrared, visible and ultraviolet. Infrared means lower than red. Hence, it stands for light frequencies that are lower than the edge of the visible red frequency. Visible red frequency starts at 430 THz or a wavelength of 700 nm. Here "T" stands for tera or 10^{12}. Infrared radiation covers frequencies' range from 300 GHz to 430 THz. That corresponds to the wavelength's range from 1 mm down to 700 nm. Here "n" stands for nano or 10^{-9}.

Visible spectrum covers the part of the electromagnetic range that our eyes could see. That starts from 430 THz to 790 THz, which is the edge of the ultraviolet. That corresponds to wavelengths from 700 nm down to 380 nm. It might be of interest to zoom in the visible spectrum and shed more light on the wavelengths of the different color components. Figure 9.8 shows an expanded visible spectrum plotted versus the wavelength in the linear scale. Based on this figure we can generate table 9.2, which classifies the wavelength/frequency ranges for each of the commonly known seven color labels.

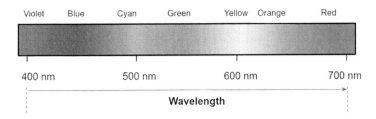

Figure 9.8: Zoom on the visible spectrum versus the wavelength of the electromagnetic radiation with some color labels.

The ultraviolet spectrum covers the frequencies range from 790 THz to 3 PHz, where "P" stands for peta or 10^{15}. That corresponds to a wavelength range from 380 nm down to 100 nm. Frequencies higher than 3 PHz are classified as ionized radiation. That includes extreme UV (from 3 PHz to 30 PHz or wavelengths from 100 nm down to 10 nm), X-rays (from 30 PHz to 30 EHz or wavelengths from 10 nm down to 10 pm) and Gamma rays (frequencies above 30 EHz). Here "E" stands for exa or 10^{18} and "p" stands for pico or 10^{-12}.

Table 9.2: Break down the visible spectrum into seven color ranges.

Color	Frequency range (THz)	Wavelength range (nm)
Violet	670 – 790	380 – 450
Blue	620 – 670	450 – 485
Cyan	600 – 620	485 – 500
Green	530 – 600	500 – 565
Yellow	510 – 530	565 – 590
Orange	480 – 510	590 – 625
Red	400 – 480	625 – 750

Sources of radiation

As for any wave, an oscillator is the source that generates such waves. Let us consider the simplified series LC oscillating circuit in figure 9.1 and examine how we can use such a model as a source of electromagnetic radiation. We know from chapter seven that the voltage drop across the capacitor is $V(t) = V_o \cdot \cos(\omega_o t)$, where $\omega_o = 1/\sqrt{LC}$. Here L is the coil inductance, and C is the capacitance value of the capacitor. We know from Kirchoff's voltage law that the current produced due to the change of the voltage across the capacitor is $I = C\frac{dV}{dt}$. Using equation 9.17 we can write that

$$I(t) = -C\omega_o V_o \cdot \sin(\omega_o t) \qquad (9.19)$$

So, what is current? In chapter seven we defined current as the rate of change of the total charges, $I = \Delta Q/\Delta t$. So, if the current is flowing in an electrical wire that has N charges per unit volume, then the total charge in a volume $\mathbf{V}$ is $Q = N\mathbf{V}q$. If the wire has a cross section A then the volume is the multiplication of the cross section by the length, l, of the segment of the wire, $\mathbf{V} = lA$. If at time t the charges are moving with a speed $u(t)$, then within a small interval Δt the charges would move a distance $l = u(t)\Delta t$ as visualized in figure 9.9.

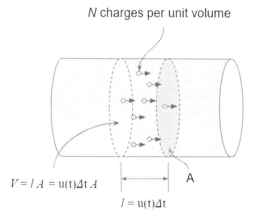

Figure 9.9: Charges flowing through a cylindrical wire form a current that is proportional to the change of changes per unit time.

From the figure one could say that all the charges inside a volume $\Delta V = Au(t)\Delta t$ will leave it during the time interval Δt. Hence, the total charge changes by the amount $\Delta Q = qN\Delta V$ during this time interval or $\Delta Q = qNAu(t)\Delta t$. In other words, the current flowing through the wire is

$$I(t) = \Delta Q/\Delta t = qNAu(t) \tag{9.20}$$

Using equations 9.20 and 9.19, one can write the speed of one charge as

$$u(t) = -\frac{CV_o\omega_o}{qNA}\sin(\omega_o t) \tag{9.21}$$

The acceleration is then the derivative of equation 9.21

$$a(t) = \frac{du(t)}{dt} = -\frac{CV_o\omega_o^2}{qNA}\cos(\omega_o t) \tag{9.22}$$

As shown, the simple LC model produces oscillating charges which in turn generate an electromagnetic wave as described in the previous section. From equation 9.11 we can write the following

$$E(t) \approx E_o \cos(\omega_o t - kr) \tag{9.23}$$

where $E_o = \frac{k_e CV_o\omega_o^2}{2NAc^2 r}$.

Example 9.2:

If an LC circuit is designed to generate radio waves in the frequency (HF) range where a $0.5\mu F$ capacitor is used, what is the needed coil inductance to ensure the generation of these waves.

Solution:

Frequency: Based on figure 9.7, the HF waves vary from 3×10^6 to 30×10^6 Hz. Or 3 MHz to 30 MHz, where M stands for "mega" or 10^6. We can select $v = 10MHz = 10 \times 10^6 Hz$.

Angular frequency: $\omega_o = 2\pi v = 6.2832 \times 10^7 rad/s$

The capacitance: $C = 0.5\mu F$

$\omega_o = 1/\sqrt{LC} \rightarrow LC = 1/\omega_o^2 = 1/(6.2832 \times 10^7)^2 = 2.53 \times 10^{-16}s$

The inductor needed is

$$L = 2.53 \times \frac{10^{-16}}{5} \times 10^{-7} = 0.506 \times 10^{-9} H = 0.506 \, nH$$

Here "n" stands for nano or 10^{-9}.

Example 9.3:

Design an LC circuit to generate radio waves in the super low frequency (SLF) range.

Solution:

Frequency: Based on figure 9.7, the SLF waves vary from 30 Hz to 300 Hz. We can select $v = 100 Hz$.

Angular frequency: $\omega_o = 2\pi v = 628.32 \, rad/s$

$$\omega_o = 1/\sqrt{LC} \rightarrow LC = 1/\omega_o^2 = 1/(628.32)^2 = 2.53 \times 10^{-6} s$$

If we select a wire inductance of $L = 5 \, mH$ then

$$C = 2.53 \times 10^{-6}/5 \times 10^{-3} = 5.07 \times 10^{-4} F = 507 \times 10^{-6} F = 507 \mu F$$

Here μ stands for micro or 10^{-6}.

What about light sources? Light frequencies are in the range of 10^{14} Hz or hundreds of THz. For that to be generated, the LC product needs to be in the range of 10^{-28}. These are small values, and one needs to think about oscillators that are on a small scale so they provide the needed charge oscillations at the desired frequency.

Light sources

When considering high frequencies and smaller time scales, one needs to think of oscillators at smaller physical dimensions, at the molecular level for instance. At his level, molecules will move, agitate or oscillate when sufficient amount of energy is given to them. One obvious source of energy that we are familiar with and use in daily bases is heat or thermal energy.

Thermal radiation

Heat provides thermal energy that increases the kinetic energy of the molecules which form the body. The molecules then start to move, rotate, vibrate or oscillate. In classical way we could say the following statement. When molecules move the bounded charges within these modules experience several forms of motion including oscillations. Part of this energy is transferred into light energy.

Thermal energy(heat) ⇒ *Kinetic energy* (molecule's motion) ⇒
Light energy (photon)

One could follow up saying that the broad range of temperature produced by heating a body of molecules would cause a large range of frequencies to be emitted. This goes well along the observation of a flame on a candle. There we clearly see an obvious change of the flame color as shown in figure 9.10. This is the same for the case of a heated metallic bar.

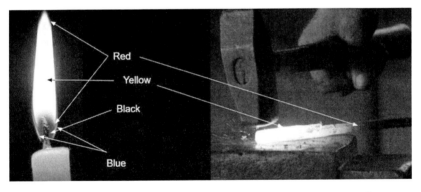

Figure 9.10: Changes of the color in the flame and heated metal rod is a reference to the change of the temperature at that region.

Heat gives thermal energy. We could in principle say that thermal energy represents the kinetic energy of the molecules of the matter. It is typically indicated by a quantity called temperature, in Kelvin (K).

$$U_{Thermal} = k_B T \qquad\qquad (9.24)$$

The constant k_B is known as Boltzmann constant and it equals $1.380649 \times 10^{-23} J \cdot K^{-1}$. Notice here that the unit of temperature is Kelvin not Celsius. This is the SI unit for temperature and it is defined as

$$T = T_C + 273.15 \qquad (9.25)$$

Where T_C is the temperature in Celsius. So, the freezing temperature in Kelvin is 273.15 K (or 0° C). The boiling temperature in Kelvin is 373.15 K (or 100° C). We will be covering more on the thermal energy and energy transfer later in thermodynamics. For now, we focus on the aspect of how heat causes electromagnetic radiation. Unfortunately, it is quite hard to find a way to explain the link between the thermal energy and frequency of oscillation/radiation without the necessity to touch some basics of quantum mechanics. For that, we will do our best to simplify the concept in order not to overwhelm students at this stage of their course of study and to keep the book within the main scope of classical mechanics.

Touching Quantum mechanics

At this stage, we can think of quantum mechanics as the movement from a continuous space into one that is built of discrete blocks and levels. We also move from the realm of defined quantities into the realm of uncertainty. However, the topic in hand does not require to dwell deep into its philosophy but rather touch the surface with careful hand in order not to lose the focus regarding our target. When dealing with energy, in quantum mechanics we assume that the amount of energy an object can take or generate is not continuous. It rather has discrete levels as shown in figure 9.11a. In the figure the index of the energy level increases proportional to the energy value where index 0 is commonly known as the ground state.

What does this mean? To understand this new way of observing the physical properties of the system, let us assume that an object such as the molecules inside the body under study have an initial energy (for example a stored potential energy) that sets them at level one. If we isolate one molecule only then we would state that it has an energy E_1 as shown in figure 9.11b. If an external source of energy is applied (such as heat) with an amount of energy that equals the difference between E_2 and E_1, $U = E_2 - E_1$, then the molecule will increase its energy and move to the energy level 2.

$$U_{final} = U_{initial} + E_2 - E_1 = E_1 + E_2 - E_1 = E_2 \qquad (9.26)$$

This mechanism when the object accepts an external amount of energy and increases its total energy is referred to as **absorption**. The molecule in our example absorbs the energy $U = E_2 - E_1$ and moves to level 2 as shown in figure 9.11c. If the applied energy cannot shift the object to one of the discrete levels, then this energy will not be absorbed as illustrated in figures 9.12a and 9.12b. In the first case a slightly lower energy is applied and hence, it was not absorbed by the object. The same happens when the energy is slightly higher than $E_2 - E_1$.

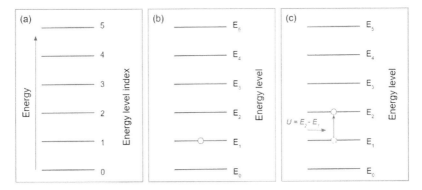

Figure 9.11: (a) Discrete energy levels (b) Original state of the body that has initial energy that equals E1. (c) Absorption of energy $U = E2 - E1$ and increase the body energy to move to level 2.

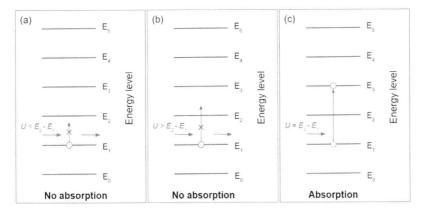

Figure 9.12: (a) The applied energy is slightly less than E2-E1 and hence no absorption. (b) The applied energy is slightly higher than E2-E1 and hence no absorption. (c) The applied energy equals E3-E1, hence the energy is absorbed, and the object shifts to level 3.

However, if the applied energy increases such that it equals $E_3 - E_1$ then the object absorbs this energy and moves to level 3 as shown in figure 9.12c. If we wish to make a certain object able to absorb more values of energy, then the difference between its energy levels needs to be reduced. Hypothetically an object that can absorb all values of energy applied to it is commonly called a **black body**.

The second mechanism in quantum mechanics is called emission. That is the opposite of absorption. Emission happens when an object such as a molecule inside the body loses energy and moves to a lower level. The body then emits an amount of energy that is equal to the difference between the two levels.

$$U_{final} = U_{initial} - \Delta E \qquad\qquad\qquad\qquad (9.27a)$$

$$U_{emit} = \Delta E \qquad\qquad\qquad\qquad\qquad (9.27b)$$

Here ΔE is the difference between the initial energy level and the final one. The difference of energy ΔE can be emitted in two forms: electromagnetic radiation or non-radiative emission (such as heat). The electromagnetic radiation in quantum mechanics is defined by a **photon**. Photon is an energy of radiation (or a light particle) that depends on the frequency of oscillation of this radiation.

$$E_{photon} = h\upsilon \qquad\qquad\qquad\qquad\qquad (9.28)$$

Where h is Planck's constant, that equals $6.62607015 \times 10^{-34} J \cdot s^{-1}$, and υ as we know is the frequency of oscillation of the emitted electromagnetic radiation. When the object shifts from energy E_3 to E_2 for instance, it will emit a photon that has a frequency of

$$\upsilon_{3\rightarrow 2} = \frac{E_3 - E_2}{h} \qquad\qquad\qquad\qquad (9.29a)$$

Similarly, if the body moves from E_3 to E_1 instead, it emits a photon with a frequency of

$$\upsilon_{3\rightarrow 1} = \frac{E_3 - E_1}{h} \qquad\qquad\qquad\qquad (3.29b)$$

One can obviously see that $\upsilon_{3\rightarrow 1} > \upsilon_{3\rightarrow 2}$. Similarly

$$\upsilon_{3\rightarrow 0} = \frac{E_3 - E_0}{h} \qquad\qquad\qquad\qquad (3.29c)$$

, where $v_{3\to0} > v_{3\to1} > v_{3\to2}$. This process of electromagnetic radiative emission is illustrated in figure 9.13.

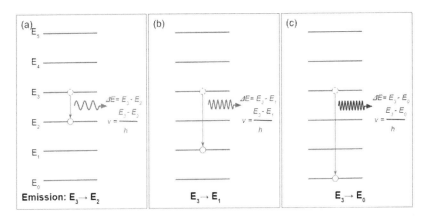

Figure 9.13: (a) The applied energy is slightly less than $E_2 - E_1$ and hence no absorption. (b) The applied energy is slightly higher than $E_2 - E_1$ and hence no absorption. (c) The applied energy equals $E_3 - E_1$, hence the energy is absorbed, and the object shifts to level 3.

In the quantized energy levels representation of the body, there are two mechanisms from which the classical physical properties can be explained: absorption and emission. To understand that, let us look at several molecules inside the body that absorbs heat energy then emits electromagnetic radiation. To represent the molecules distribution, we can place several circles at the energy level to represent how many molecules that have this value as illustrated in figure 9.14. Here, all of them are assumed to be originally at the ground state. Assume that a source of thermal energy, $U_{thermal}$, is applied. We are going to carefully select the temperature such that the applied energy equals the difference between the third. Level and the ground state, $U_{thermal} = K_B T = E_3 - E_0$. In this case and following the concept of absorption from quantum mechanics point of view, all the molecules will gain energy and move to level 3 as shown in figure 9.14b.

The molecules however will not like to keep this high energy for long as nature is lazy. There is always a tendency in nature to get rid of the excess energy. If we run for some time, we tend to walk. If we walk for a long time we tend to stand for a break. If we stand for a long time we tend to set down.

If we set for long, we tend to lay down. Basically, we are moving from high energy (kinetic and potential) to lower and lower energy levels.

Similarly for the molecules who absorbed the thermal energy, after a certain average amount of time, typically referred to as a lifetime, they will tend to move down to lower energy levels. If we assume that the energy reduction is only in radiation. This shifting of the object from high to lower energy levels generates a photon with a frequency that depends on the difference between the initial energy level and the final one as discussed earlier. This process of radiation is illustrated in figure 9.14c.

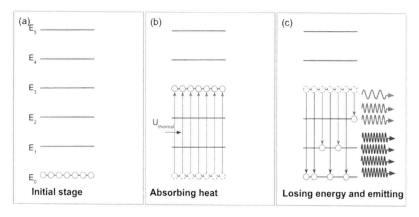

Figure 9.14: Quantized energy level representation of a group of molecules that absorbs heat and emits electromagnetic radiation at different frequencies.

As seen in the figure, the emitted electromagnetic radiation contains several frequencies depending on how much energy each particle loses. Nevertheless, the figure illustrates that there is a dominant frequency (in this example it is $v_{3\to0}$) at which most of the photons are emitted. This is commonly called the peak frequency, or v_{peaak}, and it defines in one way the dominant color of the radiation.

To examine the effect of temperature on the color of radiation, let us consider that the heating temperature is reduced so that initially all molecules are at E_2 instead. In this case we anticipate that the peak frequency will change as demonstrated in figure 9.15. As seen in figure 9.15c, lowering the applied temperature causes the peak frequency to shift to $v_{2\to0}$ instead of $v_{3\to0}$ when the higher temperature was applied. As $v_{2\to0} < v_{3\to0}$, then lowering the temperature reduces the peak frequency of the electromagnetic radiation. This in turn changes the dominant color of radiation.

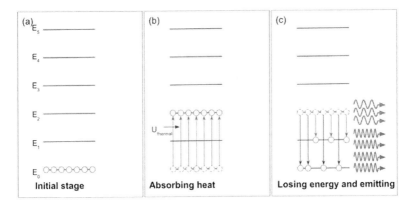

Figure 9.15: Thermal radiation when heating the particles to lower temperature.

Based on this simple discussion one can expect a proportionality between temperature, T, in Kelvin and the peak frequency, v_{peak}. This proportionality is known as Wien's displacement law.

$$v_{peak} = T \times 5.879 \times 10^{10} Hz/K \qquad (9.30)$$

In equation 9.30, the body (collection of molecules) is assumed to be a black body. In other words, it absorbs all the ranges of applied frequencies and can emit the whole range of possible frequencies. This model of thermal radiation is commonly referred to as **black body radiation**. We are not going to cover this topic in detail here, but we are going to benefit from Wien's displacement law in equation 9.30 to gain an insight on how a heated metal bar or a flame around a candle glow in different colors. To get this insight one needs to rewrite equation 3.30 using the wavelength as we are familiar with the visible light spectrum in figure 9.8 in terms of wavelength. We know that $\lambda = c/v$ where c is the speed of light. Hence, we can estimate the peak wavelength as

$$\lambda_{peak} = \frac{c}{v_{peak}} = \frac{b}{T}, b = 2.897771955 \times 10^{-3} m \cdot K \qquad (9.31)$$

The constant b is known as the Wien's displacement constant. Using equation 9.31 we can now generate a graph of the peak color as a function of the heating temperature as shown in figure 9.16.

Temperature

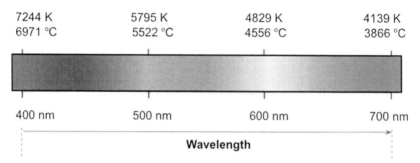

Figure 9.16: The dominant color of black body radiation as a function of the heating temperature.

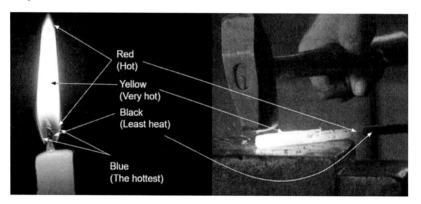

Figure 9.17: Glowing of the flame and heated metallic rod where the colors are labelled with respect to their temperature from least hot to the hottest.

If you manage to follow the simplified arguments in this section you should be able to have a conceptual understanding of why the metallic rod glows when it is heated, how the incandescent light bulb works and to some extent how the sun and other stars emit their radiation through space. From the graph in figure 9.16 we can now revisit the flame and heated rod examples and indicate how hot is it based on the emitted color as demonstrated in figure 9.17.

Example 9.4:

What is the temperature of a heated metallic rod that glows in yellow color? What would be the indication if the rod turns red?

Solution:

From table 9.2, we could guess that yellow color could be corresponding to the wavelength of 580 nm. Using Wien's displacement law in equation 9.30

$$T = \frac{b}{\lambda_{peak}} = \frac{2.897771955 \times 10^{-3} m \cdot K}{580 \times 10^{-9} m} \approx 4996\ K \approx 4723^o C$$

Red can be represented by the wavelength of 650 nm. The temperature is then

$$T = \frac{b}{\lambda_{peak}} = \frac{2.897771955 \times 10^{-3} m \cdot K}{650 \times 10^{-9} m} \approx 4458\ K \approx 4185^o C$$

The color change indicates a reduction of the temperature by $4723^o - 4185^o = 538^o$ degrees.

9C. Electricity and electromagnetic radiation

In the previous discussion, we stated that one example of thermal radiation is the incandescent light bulb. An alert student should now raise his hand and ask, "is not it an electrical light bulb?" The answer is yes, it is electrical in terms of the first source of energy, however the process of generating electromagnetic radiation is still thermal. The key work here is a known quantity that is the electrical resistance. When a current, I, flows through a resistor, it causes a potential difference of $V = IR$. We know that the electrical potential difference between the two resistor's ports is the work performed (or energy consumed) in bringing the charges, Q, across the resistor. We could in a simplified way say that

$$W = \Delta U = QV = QIR \tag{9.32}$$

If the current flowing is constant, we could say that the rate of change of the energy per unit time is

$$\frac{dW}{dt} = IR\frac{dQ}{dt} \tag{9.33}$$

We know from before that the current is the rate of change of charges per unit time, or $I = \frac{dQ}{dt}$. Similarly, the rate of change of energy per unit time is known as power, or $P = \frac{dW}{dt}$. Power has units of Watt or W for short, that is Joule per second or J/s. Using these two definitions, equation 9.33 can be rewritten as

$$P = I^2R \tag{9.34}$$

This is known as **Joule–Lenz law**. So, for a current I flowing through a resistor R, the amount of power dissipated in a form of heat is defined by equation 9.34. This heat causes the body temperature to increase, generating electromagnetic radiation as explained in the previous section. The temperature the body reaches due to the application of heat depends strongly on the material of the body.

So electrical charges which move with a constant speed form a current (DC current). This current can generate heat when passing through a resistor. This heat generates an electromagnetic radiation as we discussed earlier. The question now is, what other form of radiation the current flow could possibly generate?

Charges moving with a constant speed

In chapter seven when discussing electrical oscillators, we briefly stated that when charges move through an inductor, or coil, magnetic field lines are generated around it. The sum of all the magnetic field lines over the coil cross section was defined as the magnetic flux, Φ. This flux is linearly proportional to the current with a proportionality constant that equals the inductance L. In that discussion, we did not however explain what the nature of this magnetic field is and how it is connected to the movement of the charges. To build a better understanding, we would need to revisit Coulomb's law again. However, we will now assume two charges that move with constant speed instead of one oscillating while the other is fixed.

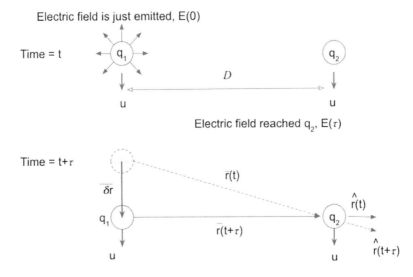

Figure 9.18: Electric field emitted by q_1 at time t reaches a moving charge q_2 at time $t + \tau$. During the propagation time, the charge q_1 moves by a displacement of $\overline{\delta r}$.

In the first section we estimated the electric field observed at a faraway location generated by an oscillating charge. The results showed that the dominant field component was oriented along the direction of oscillation of the charges and its amplitude is inversely proportional to the distance from the source. The other two field components were neglected as their amplitudes are reduced by the square and cube of the distance respectively. In our case of interest here where a current flows through a coil or a wire the field lines are to be observed in a nearby location. Hence, we cannot use the same approximation in neglecting the square and cube terms as before. However, to keep our analysis as simple as possible, we will investigate the effect of one moving charge on another when both charges move with the exact same constant velocity as shown in figure 9.18. From the figure we can see that at an instant t charge q_1 emits an electric field. After a time delay of τ, the field reaches charge q_2 and exerts a force on it. In comparison to the first case in figure 9.5, here q_2 has moved with the exact same velocity as q_1 during the time delay. The distance the electric field travels to reach q_2 in this case is larger than the separation from the charges after the time delay, $r(t) > r(t + \tau)$, as visualized in figure 9.18. From equations 9.4 and 9.5 we know that the force that is experienced on charge q_2 is

$$\overline{F}(t + \tau) = q_2 \frac{k_e q_1}{r^2(t)} \hat{r}(t) \tag{9.35}$$

Both charges are moving with the exact constant velocity along the y direction, for example, $\overline{u} = u\hat{y}$, and they are separated by a distance D along the x direction as illustrated in the figure. The separation D is not large and hence the time delay τ is sufficiently small such that radial component of the electric field is dominant, $|\overline{F}| \approx F_r$. Hence, we can approximately write the force as

$$\overline{F}(t + \tau) = q_2 \frac{k_e q_1}{r^2(t)} \hat{r} \tag{9.36}$$

We know that the distance $r(t)$ is larger than $r(t + \tau)$. Hence the force between the two moving charges is less than that which would be caused by two static charges at the same separation distance. Knowing that the distance $\delta r = u\tau$, one can write the following relation using the geometry in figure 9.18,

$$r^2(t) = r^2(t + \tau) + u^2\tau^2 \tag{9.37}$$

As discussed earlier, we assumed that the electric field travels at the speed of light and

hence the delay $\tau = r(t)/c$. Using that in equation 9.37 we obtain

$$r^2(t) = r^2(t + \tau)^2 + \frac{u^2 r^2(t)}{c^2} \tag{9.38}$$

Rearranging equation 9.38 we obtain the following relation

$$r(t) = \frac{r(t+\tau)}{\sqrt{1-(u/c)^2}} = \gamma r(t + \tau) \tag{9.39}$$

Notice that the factor $\gamma = 1/\sqrt{1 - u^2/c^2}$ is known as the **Lorentz factor**, which describes how the physical properties of an object (such as time and dimension) change when it moves. It is an important factor in Einstein's special relativity. Substituting equation 9.39 into 9.36 we obtain

$$\overline{F}(t + \tau) = q_2 \frac{k_e q_1}{r^2_{(t+\tau)}} \hat{r} - q_2 \frac{k_e q_1 u^2}{r^2_{(t+\tau)}} \hat{r} \qquad (9.40a)$$

As both force and displacement in Equation 9.40a are at the same time frame, then we can remove the time dependence in the equation and set $r = D$ (the separation between the charges)

$$\overline{F} = q_2 \frac{k_e q_1}{D^2} \hat{r} - q_2 \frac{k_e q_1 u^2}{D^2} \hat{r} \qquad (9.40b)$$

The first term in equation 9.40b represents Coulomb's force for two static charges separated by a distance D, this is referred to as the electrostatic force or $\overline{F}_E$ in short, where $\overline{F}_E = q_2 \frac{k_e q_1}{D^2} \hat{r}$. The second term however depends mainly on the speed of the two charges. This is commonly known as the magnetostatic force or $\overline{F}_B$ in short, where $\overline{F}_B = -q_2 \frac{k_e q_1 u^2}{D^2} \hat{r}$. So, the total force is $\overline{F} = \overline{F}_E + \overline{F}_B$. We can see that if both chargers have the same sign, then the electrostatic force will be pushing them apart. However, the magnetostatic force will be acting on pulling them together. That is why the total force is reduced. In other words, when both particles move at the same speed, their motion causes the force in between them to reduce due to the rise of the magnetostatic force that is generated by this motion.

Magnetostatic force due current flow in a wire

When considering a flow of charges in a metallic wire, one can assume the scenario in figure 9.19. There, the wire has a cross-section area A and length l that is much larger than the cross-section dimensions and the separation D. As illustrated in the figure, at time instant t, there exists large number of charges inside the wire. It would be a tedious job to consider the effect of every charge individually and add up the total force. Not to mention that we do not know for certain the location of each charge within the volume of the wire at that instant. To resolve this difficulty, we can assume a very small segment of the wire that has a very small length dy and cross section area A as highlighted by the dark desk in figure 9.19. The segment can be then dealt with in a similar way as one charge element where the charge amplitude

equals the sum of all charges inside the volume of this segment. If the wire
has a known density of N charges per unit volume, then we can estimate the
charge inside the small segment as

$$Segment\ charg = Density \times segment\ Volume \\ \times individual\ Charge\ amplitude$$

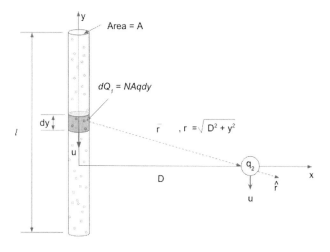

Figure 9.19: The effect of a large number of charges flowing through a wire on a
charge q2 the moves with the same velocity as those inside the wire.

From the geometry in figure 9.18, we can state that the small segment
volume is the multiplication of the cross-section area, A, by the segment
length dy, or $dV = Ady$.

If every charge inside the wire has an amplitude q, then the charge of the
segment is

$$dQ_1 = N \times Ady \times q \tag{9.41}$$

Replacing q_1 in equation 9.40b with dQ_1 the force on q_2 due to only the
small segment is

$$d\overline{F}_B = -k_e \frac{uq_2 \cdot u\,dQ_1}{c^2 r^2} \hat{r} \tag{9.42}$$

Notice that we expand the term u^2 as a multiplication of $u \cdot u$ in equation 9.42. Replacing dQ_1 by $NAqdy$ and r by $\sqrt{D^2 + y^2}$ we obtain

$$d\overline{F}_B = -k_e \frac{uq_2 \cdot uNAq}{c^2} \frac{dy}{D^2 + y^2} \hat{r} \tag{9.44}$$

From the previous discussion on current relation to moving charges in equation 9.20 we know that the current in the wire is $I_1 = u \cdot NqA$. Using this expression, we can arrange equation 9.44 to obtain the force per unit length of the wire on charge q_2 as

$$\frac{d\overline{F}_B}{dy} = -k_e \frac{uq_2 \cdot I_1}{c^2} \frac{1}{D^2 + y^2} \hat{r} \tag{9.45}$$

Magnetostatic force between two currents

In the previous section we examined the effect of a current flowing through a long conducting wire over a charge element that is moving parallel to the wire with the exact same velocity as the charges inside the conductor. That resulted in a force per unit length of the wire as expressed in equation 9.45. Now we should be ready to examine the more practical case of the force exerted on a current flow in one wire due to the flow of another current in a parallel wire as visualized in figure 9.20.

Similar to the way we handled the large number of charges moving through the first wire, we can now replace the charge element q_2 by the total charge, dQ_2, inside a small segment of length dy'. When assuming that the second wire has a density N_2 charges per unit volume, then $dQ_2 = NqAdy'$ and equation 9.44 becomes

$$d(d\overline{F}_B) = -k_e \frac{uN_2qAdy' \cdot I_1}{c^2} \frac{dy}{D^2 + y^2} \hat{r} \tag{9.46}$$

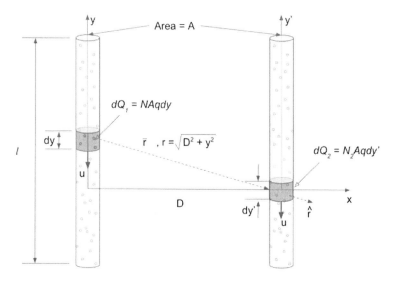

Figure 9.20: The effect of a large number of charges flowing through a wire on a current flowing through a parallel wire.

Notice that here we use the double derivative $d(d\bar{F}_B)$ to stress that it is the effect of the small segment of the wire dQ_1 on the small segment dQ_2. Again, we can replace $u \cdot N_2qA$ by the current I_2 that flows through the second wire and dividing both sides by dy',

$$d\left(\frac{d\bar{F}_B}{dy'}\right) = -k_e \frac{I_2 \cdot I_1}{c^2} \frac{dy}{D^2+y^2} \hat{r} \qquad (9.47)$$

From the geometry in figure 9.20, we can write the unit vector $\hat{r} = \frac{D\,\hat{x} + y\,\hat{y}}{\sqrt{D^2+y^2}}$. Hence, the force exerted per unit length on wire 2 due to the current in wire 1, $\frac{d\bar{F}_B}{dy'}$, as the integration of equation 9.47

$$\frac{d\bar{F}_B}{dy'} = -k_e \frac{I_2 \cdot I_1}{c^2} \int_{y=-\infty}^{\infty} \frac{dy}{D^2+y^2} \hat{r} \qquad (9.48)$$

Here, we assumed that the write has an infinite extension in both directions. We can now apply the expansion of the unit vector into the equation as

$$\frac{d\bar{F}_B}{dy'} = -k_e \frac{I_2 \cdot I_1}{c^2} \int_{y=-\infty}^{\infty} \frac{D\,\hat{x} + y\,\hat{y}}{(D^2+y^2)^{3/2}} dy \qquad (9.49)$$

We can write equation 9.49 as two scalar equations, one along the x axis and the other along the y axis.

$$\frac{dF_{B,x}}{dy'} = -k_e \frac{l_2 \cdot l_1}{c^2} \int_{y=-\infty}^{\infty} \frac{D}{(D^2+y^2)^{3/2}} \, dy \tag{9.50a}$$

$$\frac{dF_{B,y}}{dy'} = -k_e \frac{l_2 \cdot l_1}{c^2} \int_{y=-\infty}^{\infty} \frac{y}{(D^2+y^2)^{3/2}} \, dy \tag{9.50b}$$

As can be proven the integration 9.50b is over an odd function, $f(-y) = -f(y)$. Hence the y component of the force per unit length is zero, $\frac{dF_{B,y}}{dy'} = 0$. For the x component, we use the following change of variables:

$$y = D \cdot tan\theta \tag{9.51a}$$

$$dy = \frac{D}{cos^2\theta} d\theta \tag{9.51b}$$

Then

$$\frac{dF_{B,x}}{dy'} = -k_e \frac{l_2 \cdot l_1}{c^2} \int_{\theta=-\pi/2}^{\pi/2} \frac{D}{D^3(1+tan^2\theta)^{3/2}} \left(\frac{D}{cos^2\theta} d\theta \right) \tag{9.52a}$$

$$= -k_e \frac{l_2 \cdot l_1}{c^2} \int_{-\pi/2}^{\pi/2} \frac{cos^3\theta}{D} \left(\frac{d\theta}{cos^2\theta} \right) = -k_e \frac{l_2 \cdot l_1}{c^2 D} \int_{-\pi/2}^{\pi/2} cos\,\theta d\theta \tag{9.52b}$$

$$= -k_e \frac{l_2 \cdot l_1}{c^2 D} sin\,\theta \, |_{-\pi/2}^{\pi/2} = -2k_e \frac{l_2 \cdot l_1}{c^2 D} \tag{9.52c}$$

The result in equation 9.52c indicates the force between two parallel wires is inverse proportional to the separation between them, D, and points along the x-axis (direction normal to both wires). This is also known as Ampere's force law. So, the magnitude of the force of attraction or rebelling between two parallel currents of length l can be approximately estimated in a vector form as follows

$$d\overline{F}_B = -\frac{k_e l_1 l_2}{c^2 D} dy' \hat{x} \tag{9.53}$$

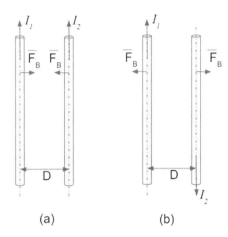

Figure 9.21: Magnetostatic force between two parallel wires when the currents flow (a) in the same direction and (b) in opposite directions.

The magnetostatic force on the wire is the integration over y' from 0 to l.

$$\overline{F}_B = -\frac{2k_e I_1 I_2}{c^2 D} \int_0^l dy' \hat{x} = -\frac{2k_e I_1 I_2 l}{c^2 D} \hat{x} \qquad (9.54)$$

Let us accept, for now, the flowing definition: the current is positive when it flows along the y axis (upwards as in figure 9.21) and it is negative when it flows along the negative y direction (downwards in figure 9.21). Hence, according to equation 9.54, when the two currents flow in the same direction their product is positive and the force between them is attractive as in figure 9.21a. Rebelling force is obtained when the two currents flow at opposite directions as illustrated in figure 9.21b.

Rise of the magnetic field

The magnitude of the magnetostatic force between the two parallel wires is

$$F_B = |\overline{F}_B| = \frac{2k_e I_1 I_2 l}{c^2 D} \qquad (9.55)$$

We can rewrite the magnitude of the force as

$$F_B = I_2 \cdot \left(\frac{2k_e I_1}{c^2 D}\right) \cdot l \qquad (9.56)$$

The term in parentheses will be referred to as the magnitude of the magnetic field

$$B_1 = \frac{2k_e I_1}{c^2 D} \qquad (9.57)$$

Equation 9.56 now becomes

$$F_B = I_2 \cdot B_1 \cdot l \qquad (9.58)$$

In other words, we can say the current I_1 generates a magnetic field, B_1. The field propagates for a distance D and induces a force F_B on a second wire of length l that is carrying a current I_2. As can be observed from equation 9.58 that the units of the magnetic field are Newton per Ampere per meter, N/A.m, or as commonly known as Tesla. It is as well noticeable from equation 9.57 that the magnitude of the magnetic field should be constant along a circle of radius D drawn around the wire. If the wire length is very long then we could say that the magnitude of the magnetic field is constant on the side of a cylinder of radius D drawn around the wire. This indicates cylindrical symmetry of B_1 as depicted by the cylindrical coordinates in figure 9.22a. Here I_1 is flowing along the z axis.

At any point of the cylindrical surface of radius r, the magnetic field amplitude is constant, $B_1 = 2k_e I_1/c^2 r$. If a second wire with a current I_2 is placed parallel at any location on cylinder's surface, then it will experience the same magnitude of the force per unit length $F_B/l = I_2 \cdot B_1$ compared to any other location on the same surface. When the two currents are flowing at the same direction, then the resultant force is attractive as illustrated in figure 9.22b. That means that the direction of F_B is towards the center or the negative radial direction

$$\overline{F}_B/l = I_2 \cdot B_1(-\hat{e}_r) \qquad (9.59)$$

Here, $\hat{e}_r$ is a unit vector in the radial direction pointing from the origin outwards. We know that current flows at a specific direction and hence it is actually a vector quantity that in our example can be written as $\overline{I}_2 = I_2\hat{z}$ where $\hat{z}$ is a unit vector along the z direction. In equation 9.59 we know that both current $\overline{I}_2$ and the force, $\overline{F}_B$, are vector quantities. The two vectors however point in two orthogonal directions. Hence, the magnetic field cannot be a scalar quantity. It has to be a vector quantity as well. The multiplication operator between the current and the magnetic field hence needs to be one of the two vector multiplications: dot or cross products. Dot product as we mentioned in chapter one results in a scalar quantity and hence it cannot be the proper operator. That leaves us with the cross product between the two vectors I_2 and B_1 or

$$\frac{\overline{F_B}}{l} = \overline{I}_2 \times \overline{B}_1 \tag{9.60}$$

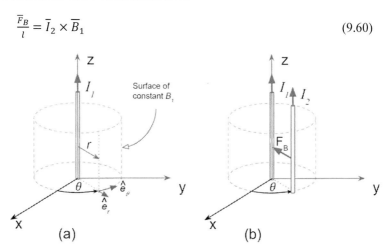

Figure 9.22: (a) Cylindrical coordinate system around the first wire where the current flows along the positive z direction. (b) Attraction force induced on a second placed parallel to $I1$ at a distance r.

The question now is which direction could the magnetic field be pointing such that the resultant force vector is pointing at $-\hat{e}_r$. There are three possibilities in the considered cylindrical coordinates system: z, r and θ as illustrated in figure 9.23a. If B_1 is along the z direction ($\overline{B}_1 = B_1\hat{z}$) then the cross product in 9.60 will be

$$\frac{\overline{F_B}}{l} = I_2 B_1 \hat{z} \times \hat{z} = 0 \tag{9.61}$$

This is of course not the desired result and hence the magnetic field cannot be parallel to the current. Now, if the field is along the radial direction ($\overline{B}_1 = B_1\hat{e}_r$) then the cross product becomes

$$\frac{\overline{F_B}}{l} = I_2 B_1 \hat{z} \times \hat{e}_r = I_2 B_1 \hat{e}_\theta \tag{9.62}$$

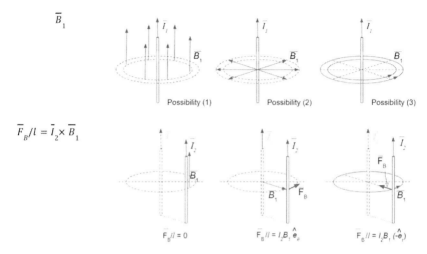

Figure 9.23: (a) Cylindrical coordinate system around the first wire where the current flows along the positive z direction. (b) Attraction force induced on a second placed parallel to $I1$ at a distance r.

The direction of the produced force is not what we obtained earlier in equation 9.59 and hence the magnetic field cannot be pointing at the radial direction. Only one possibility left which is the case when the magnetic field is pointing in the circular direction ($\overline{B}_1 = B_1 \hat{e}_\theta$). In this case

$$\frac{\overline{F}_B}{l} = I_2 B_1 \hat{z} \times \hat{e}_\theta = -I_2 B_2 \hat{e}_r \tag{9.63}$$

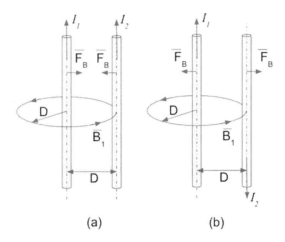

(a)   (b)

Figure 9.25: The force between two currents flowing in parallel wires when the currents are (a) along the same direction and (b) at opposite directions.

This result matches our earlier derivation. The three possibilities of the magnetic field and the resultant force are illustrated in figure 9.24. So, when a current flows along the z direction a magnetic field is produced in a circular direction centered around the current flow. When another wire carrying a current is placed nearby, it experiences a force per unit length that is the cross product of that current and the present magnetic field. We can correct the representation in figure 9.21 as in figure 9.25.

Electromagnetic force

We can generally say the magnetostatic force on a current, I, due to the presence of a magnetic field B over a length l is

$$\overline{F}_B = (\overline{I} \times \overline{B})l \tag{9.64}$$

Using the definition of current from equation 9.20, we can obtain the following

$$\overline{F}_B = (NAq\overline{u} \times \overline{B})l \tag{9.65a}$$

$$\overline{F}_B = q(NAl)\overline{u} \times \overline{B} \tag{9.65b}$$

Here Al is the volume of the wire where the current flow and N is the number of charges per unit volume. If we assume that the current is very small such that only one charge is present at a certain instant of time, then the produce NAl should equal one and equation 9.65b for the magnetic force that is applied on the charge that is moving with velocity $\overline{u}$ is

$$\overline{F}_B = q\overline{u} \times \overline{B} \tag{9.66}$$

If an electric field $\overline{E}$ is present, we know that it as well exerts an electrostatic force on the charge following equation 9.6, $\overline{F}_e = q\overline{E}$. The total force that is then applied on the charge is the summation of both forces, electrostatic and magnetostatic

$$\overline{F} = q\overline{E} + q\overline{u} \times \overline{B} \tag{9.67}$$

This is known as the electromagnetic force or Lorentz force.

Example 9.5:

An electron is emitted from an electron gun with a velocity of $10^6 \, \hat{x} \, m/s$. It passes through a capacitor of 2 cm width and 2 cm spacing. The capacitor has an applied potential difference of 5 V between the to plates. If we ignore the relativistic effects, calculate the velocity of the electron when it exits the capacitor.

Solution:

As we have learned in chapter seven, when applying a potential difference V between the capacitor plates, an electric field is generated inside the capacitor where the filed lines point from the positive to the negative sides. The strength of the electric filed is V/d where d is the spacing between the plates. Hence, we can write the produced electric field as

$$\overline{E} = -\frac{V}{d}\,\hat{y} = -\frac{5V}{0.02m}y = -250\,\hat{y}\,V/m$$

The force that would be applied on the electron when it is between the plates has a strength of qE and it is pointing downwards along the negative y direction.

$$\bar{F} = -qE\,\hat{y} = -1.602 \times 10^{-19} C \times 250\frac{V}{m}\,\hat{y} \approx -4 \times 10^{-17}\,N\,\hat{y}$$

If we apply Newton's second law of motion, the electron will experience a constant acceleration when traveling between the capacitor plates that is

$$\bar{a} = \frac{\bar{F}}{m} = -\frac{4 \times 10^{-17}}{9.109 \times 10^{-31}}\,\hat{y} \approx -4.4 \times 10^{13}\frac{m}{s^2}\,\hat{y}$$

The velocity is the integration of the acceleration over time

$$\bar{u} = \bar{u}_o + \int \bar{a}\,dt = 10^6\,\hat{x} - 4.4 \times 10^{13}t\,\hat{y}$$

Similarly, the displacement vector is the integration of the velocity. If we assume that the electron was at the origin of the coordinates when entering the capacitor, $\bar{r}_o = 0$, then

$$\bar{r} = \bar{r}_o + \int \bar{u}\,dt = 10^6 t\,\hat{x} - 2.2 \times 10^{13}t^2\,\hat{y}$$

The time it takes for the electron to cross the capacitor is when x=2 cm. 0.02 m. Hence, the time needed for that is

$$t = \frac{0.02m}{10^6 m/s} = 2 \times 10^{-8} s$$

During this time, the y component of the velocity vector becomes

$$u_y = -4.4 \times 10^{13}t = -4.4 \times 10^{13} \times 2 \times 10^{-8} = -8.8 \times 10^5\,m/s$$

The electron then exits with a velocity of

$$\bar{u} = 10^6\,\hat{x} - 8.8 \times 10^5\,\hat{y}\,m/s$$

That makes an angle of $\tan^{-1}\left(\frac{-8.8 \times 10^5}{10^6}\right) = -41.3^o$ with respect to the x-axis.

The location of the electron along the y axis is

$$y = -2.2 \times 10^{13}t^2 = -2.2 \times 10^{13}(2 \times 10^{-8})^2 = -8.8 \times 10^{-3}m$$
$$= -8.8mm$$

9D. Summary

- The force between two static charges, q_1 and q_2 that are separated by a distance r is described by an experimental law, Coulomb's law.
  - $\overline{F} = q_2 \left(\frac{k_e q_1}{r^2}\right) \hat{r}$
  - k_e is the Coulomb's constant $\approx 8.988 \times 10^9 N \cdot m^2 \cdot C^{-2}$
  - $\hat{r}$ is a unit vector pointing from q_1 to q_2
    - If q_1 and q_2 have the same sign:
      - The force is along $\hat{r}$
      - Rebelling force
    - If q_1 and q_2 have opposite signs:
      - The force is along $-\hat{r}$
      - Attraction force
- The effect of one charge q_1 on a remote charge q_2 can be described as
  - q_1 emits an electric field: $\overline{E} = \left(\frac{k_e q_1}{r^2}\right)\hat{r}$
  - The electric field travels a distance r to q_2
  - The electric field then induces a force on q_2
    - $\overline{F} = q_2\overline{E}$
    - The direction of the force is along the electric field or opposite to it depending on the sign of the charge q_2.
    - The electric field has units of N/C or V/m
- When charge q_1 starts to oscillate, its location changes with time
  - At time t
    - q_1 is at location $r(t)$
    - q_1 emits an electric field
  - At time $t + \tau$ the electric field reaches the charge q_2
    - q_1 is now
      - At location $r(t + \tau)$
      - Direction along $\hat{r}(t + \tau)$
    - The field that reached q_2 is however
      - $\overline{E}(t + \tau) = \frac{k_e q_1}{r(t)^2}\hat{r}(t)$
      - Direction along $\hat{r}(t)$
      - It travelled at the speed of light c
        - The delay is $\tau = r(t)/c$

- Using Taylor's expansion:
  - The displacement vector can be written as
    - $\overline{r}(t) \approx \overline{r}(t + \tau) - \frac{r(t)}{c} \cdot \overline{u}(t) - \frac{r(t)^2}{2c^2} \cdot \overline{a}(t)$
  - The electric field becomes
    - $\overline{E}(t + \tau) = \frac{k_e q_1}{r^3}\overline{r}(t + \tau) - \frac{k_e q_1}{r^2 c}\overline{u}(t) - \frac{k_e q_1}{2rc^2}\overline{a}(t)$
  - When q_2 becomes very far we can write the electric field as
    - $\overline{E}(t + \tau) \approx -\frac{k_e q_1}{rc^2}\overline{a}(t)$
      - The electric field amplitude is reduced by 1/r
      - The electric field is along the direction of oscillation (acceleration) of the charge.
  - If $\overline{a}(t) = a_o \cos \omega t$ then
    - $\overline{E}(t + \tau) \approx E_o \cos(\omega t - kr)$
    - $E_o = \frac{A}{r}$ and $A = \frac{k_e q_1 a_o}{2c^2}$
    - $k = \frac{\omega}{c}$, is referred to as the wavenumber
    - $\omega = 2\pi v$ is the angular frequency
    - The wavelength is $\lambda = c/v$
  - The produced electric field is a wave that oscillates with frequency v and propagates with the speed of light.
  - This wave is typically referred to as electromagnetic wave.
- Electromagnetic waves
  - Produced by oscillating charges (classical mechanics).
  - Do not require a medium to propagate
  - Travel with the phase velocity that equals the speed of light
  - Cover frequency ranges of
    - Radio waves: Few Hertz to MHz
    - Microwaves: GHz
    - Light: THz
    - Ionized radiation: PHz and larger

- Frequencies lower than light could be in an ideal case produced by classical electrical oscillators (LC circuit for example).
  - Charges that form the electrical current oscillates between the cycles with acceleration
    - $a(t) = \frac{du(t)}{dt} = -\frac{CV_o\omega_o^2}{qNA}\cos(\omega_o t)$
    - $\omega_o = 1/\sqrt{LC}$
    - A is the area of the wire
    - N is the number of charges per unit volume.
  - The produced electric field is
    - $E(t) \approx E_o \cos(\omega_o t - kr)$
    - $E_o = \frac{k_e C V_o \omega_o^2}{2NAc^2 r}$
- Understanding light sources however requires some understanding of quantum mechanics.
- Simplified basics of quantum mechanics
  - Physical quantities are discrete rather than continuous.
    - For example, the energy values that an object can absorb are not continues.
  - We cannot be certain of the values of all the physical quantities at once.
    - If we know the object location, we are not certain of its momentum.
  - We would need to talk about possibilities rather than defined values.
    - The probability distribution of the object location instead of a defined location.
- Light sources and quantum mechanics
  - Energy absorption by an object happens in discrete nature.
    - Each object has its own discrete energy levels E_n, where n is an integer.
      - These are defined by the form the energy can be transferred to:
        - Electron excitation
        - Molecular vibration
        - Molecular rotation
        - Molecular translation.

- Consider levels E_n and E_m, where $E_m > E_n$ and incident energy U:
  - Absorption: $U = E_m - E_n$
  - No absorption: $U \neq E_m - E_n$
- Energy emission by an object
  - Objects tend to move to lower energy levels.
  - After a certain time, life time, the object moves from higher energy level E_m to a lower one E_n.
  - The difference of the energy, $E_m - E_n$, is emitted in one of two forms:
    - Radiative electromagnetic wave (Photon)
      - $h\nu = E_m - E_n$
      - Plank's constant: $h = 6.62607015 \times 10^{-34} J \cdot s^{-1}$
    - Nonradiative emission
      - Heat
- Thermal radiation
  - Heating the object
    - Excites the molecules to higher energy levels.
  - By time, many molecules move to lower energy levels.
  - Transition between different high and low energy levels generate photons of different frequencies (colors).
  - If the body is assumed to be black body:
    - Absorbs all energy levels
    - Can emit all energy levels.
    - The dominant color depends on temperature
      - Wien's displacement law
      - $\nu_{peak} = T \times (5.879 \times 10^{10} Hz/K)$
      - $\lambda_{peak} = \frac{(2.897771955 \times 10^{-3} m \cdot K)}{T}$
- Current and magnetic force
  - When a charge moves with constant speed
    - $\bar{u}(t) = u$
    - $\bar{a}(t) = 0$

- When two charges q_1 and q_2 move in parallel (separated by distance d) with the same speed u
  - The electric field emitted by q_1 travels a distance r that is longer than d to reach q_2
    - $r = \sqrt{d^2 + (u\tau)^2}$
    - The delay $\tau = r/c$
      - $r = d/\sqrt{1 - (u/c)^2}$
  - The force produced by q_1 on q_2 is
    - $\overline{F} = q_2 \frac{k_e q_1}{d^2}\hat{r} - q_2 \frac{k_e q_1 u^2}{d^2}\hat{r}$
    - Electrostatic force: $\overline{F}_E = q_2 \frac{k_e q_1}{d^2}\hat{r}$
    - Magnetic force: $\overline{F}_B = -q_2 \frac{k_e q_1 u^2}{d^2}\hat{r}$
  - When current flows in two parallel wires, the unit charges are replaced with elementary charges
    - $dQ_1 = N_1 \times A_1 dy \times q$
      - N_1: Number of charges per unit volume
      - A_1: Area of wire 1.
    - $dQ_2 = N_2 \times A_2 dy \times q$
  - The current in each wire
    - $I_1 = \frac{dQ_1}{dt} = N_1 \times A_1 \frac{dy}{dt} \times q = N_1 A_1 u q$
    - $I_2 = \frac{dQ_2}{dt} = N_2 \times A_2 \frac{dy}{dt} \times q = N_2 A_2 u q$
  - The force per length of the wire, l, is
    - $\overline{F}_B/l = -\frac{2 k_e I_1 I_2}{c^2 D}\hat{x}$
- Magnetic field
  - The magnitude of the force
    - $F_B = I_2 \cdot \left(\frac{2 k_e I_1}{c^2 D}\right) \cdot l$
    - $F_B = I_2 \cdot B_1 \cdot l$
      - $B_1 = \frac{2 k_e I_1}{c^2 D}$
      - It is the magnetic field emitted by the current I_1 flowing through wire 1.
      - The magnetic field points in a circular direction around the wire following the right-hand rule.
    - $\overline{F}_B = (\overline{I} \times \overline{B})l$
  - The electromagnetic force produced by electric and magnetic fields on charge q is then the summation of the electric and magnetic forces
    - $\overline{F} = q\overline{E} + q\overline{u} \times \overline{B}$

9E. Review questions

1. Five charges of 1 C each are placed on the y axis as in the figure. That charges are separated by 10 mm each. What is the total electric field produced by these charges at a point that is 2 mm on the x axis? *hint: Use Coulomb's law and calculate the vectorial electric field due to each point then add all the fields together.*

2. Place a label with the name of radiation for the given frequencies:

   a) 45 Hz   b) 9 MHz   c) 400 GHz   d) 600 THz

3. If possible, design an LC circuit to generate the frequencies in question 2. If the frequency is for light emission, calculate the difference between the two energy levels that produces a photon with that frequency.

4. Two fixed charges of 3 C and 4 C amplitude are placed at a distance of 0.4 mm apart.

   a. What is the amplitude and dir3ction of the force felt by each particle?
   b. What is the amplitude of the electric fields produced by each particle at a point located in the middle between the two charges?

5. There are two identical electrical wires separated by a distance of 5 mm.

   a. If a current of 400 mA is passing through each wire in the same direction, what is the force per unit length that is felt on each wire? Was it attraction or rebelling?
   b. If the direction of the second wire is reversed, how does this affect the force?
   c. If the first wire was fixed and the second wire has a length of 4 cm, thickness of 0.3 mm, and mass density of 9 g/cm3, what is the amount of acceleration it will experience for the case above?

6. A wire of length 50 cm is carrying a current of 2 A. It is placed perpendicular to a magnetic field of 0.5 T. What is the magnitude and direction of the magnetic force experienced by the wire?

7. An electrical current is formed by electrons moving in a copper wire of 0.2 mm radius at a uniform speed of 2x10^-5 m/sec. For copper, the typical density of free electrons is about 8.5x10^28 electrons/m^3. Calculate the magnetic field produced by the wire at a distance of 3 mm from the wire.

8. A capacitor composed of two metal plates separated by a distance of 3 mm. An electric potential difference of 5 V is applied across the capacitor.

    a. What is the amplitude and direction of the electric field produced by the capacitor?

    b. What is the amplitude and direction of the force that would be felt by a free charge of amplitude 2x10^-12 C that is placed in between the plated?

    c. How much time would it takes a free charge to travel from the positive plate to negative one?

CHAPTER TEN

INTRODUCTION TO FLUID MECHANICS

10A. Defining fluids

By definition, fluid is a substance that deforms continuously under the application of external shear stress. Based on our knowledge up to this point, we could understand most of the previous sentence except for the last two words: shear stress. To understand shear stress, let us consider a body made of a substance that is bounded between two rigid surfaces with a separation h in the y direction as in figure 10.1a. If we manage to pull the top surface along the x-direction at a constant force, F, then the shear stress is defined as the force per unit area of the surface. If the force is assumed constant all over the top layer of the substance and that the top surface has an area A, then the shear stress is simply

$$\tau = \frac{F}{A} \; N/m^2 \tag{10.1}$$

Shear stress has units of Newton per meter square or N/m^2 or Pascal, Pa for short. The shear stress causes a deformation of the substance as illustrated in Figure 10.1b after a certain time of application of the force.

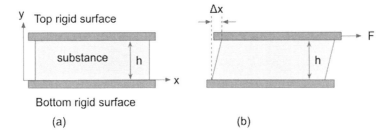

Figure 10.1: (a) a body made of substance is held between two rigid surfaces. (b) deformation of the substance due to the application of a parallel force on the top surface.

The deformation of the substance can be evaluated by the amount of displacement along the direction of the force application. This brings forward another quantity, which is the shear strain. Shear strain is defined as the displacement along the direction of the applied force normalized to the height of the body, or

$$\gamma = \frac{\Delta x}{h} \tag{10.2}$$

As a normalized quantity, shear strain does not have a unit of its own. It is typically measured in radians.

Solid and fluid

So, what is the difference between solid and fluid when shear stress is applied. According to the definition of fluids, they would experience constant deformation as long as the shear stress is applied. We could then correctly guess that solids on the other hand will stop getting deformed after a short time and remain with a constant deformation as long as the force is applied. Figure 10.1b represents the deformation in a solid object under constant force at any instant of time. Based on this we could state the following, a deformation in the form of a fixed displacement in the object occurs as long as a constant force is continuously applied. Mathematically the statement is written as

$$\Delta x \propto F \tag{10.3a}$$

If we divide both sides of the proportionality in 10.3a by the solid volume before stress, $h \cdot A$, we obtain

$$\frac{\Delta x}{h \cdot A} \propto \frac{F}{h \cdot A} \rightarrow \left(\frac{\Delta x}{h}\right)\frac{1}{A} \propto \left(\frac{F}{A}\right)\frac{1}{h} \tag{10.3b}$$

Using the definition of strain and sheer stress, we can re-write the proportionality in 10.3b as

$$\gamma \frac{1}{A} \propto \tau \frac{1}{h} \rightarrow \gamma \propto \tau \tag{10.3c}$$

Equation 10.3C states that the shear stress is proportional to the shear strain for solids. We can write the proportionality in a form of equality as

$$\tau = G\gamma \tag{10.4}$$

Where G is sheer modulus of the solid. This quantity has the same units as shear stress (N/m2 or Pa) and it is expected to be different for different solids. Table 10.1 shows some values of the shear modulus for different known solids.

Table 10.1: Shear modulus for different materials.

Materials	Shear modulus G in GPa $(10^9 \times Pa)$
Steel	79.3
Iron	52.5
Copper	44.7
Glass	26.2
Aluminum	25.5
Wood	4

As seen in the table the shear modulus for steel is higher than that of glass. What does that mean? It would simply mean that to achieve the same deformation in both substances, a much larger shear stress is needed for steel compared to glass. To get a better understanding, let us benefit from the following example.

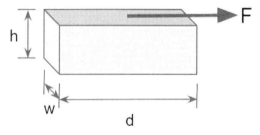

Figure 10.3: Deformation of a fluid body under constant shear stress as time progresses.

Example 10.1:

Consider the rod geometry shown in figure 10.2. The rod has a rectangular cross section of width w and height h. The length of the rod is d. A force F is applied on the top surface along the x-direction. If the dimensions are:

w = 4 *cm*, h = 10 *cm*, d = 40 *cm*

What would be the amount of displacement Δx when a constant force of 10 N is continuously applied for a substance made of iron? What would be the displacement if the rod is made of glass?

Solution:

To answer these questions, we would need to first estimate the applied shear stress. The top area is

$$A = w \times d = 0.04m \times 0.4m = 0.016m^2$$

The shear stress is

$$\tau = \frac{F}{A} = \frac{10N}{0.016m^2} = 625Pa$$

The shear strain is related to the shear stress by $\gamma = \frac{\tau}{G}$. The shear modulus of iron from

table 10.1 is 52.5 GPa or $52.5 \times 10^9 Pa$.

$$\gamma = \frac{\tau}{G} = \frac{625Pa}{52.5 \times 10^9 Pa} = 11.9 \times 10^{-9}$$

The shear strain is defined as $\gamma = \frac{\Delta x}{h}$, hence

$\Delta x = \gamma \times h = 11.9 \times 10^{-9} \times 0.1m = 1.19 \times 10^{-9}m$ or   1.19   nm displacement.
For the case of glass, $G = 25.5\,GPa = 25.5 \times 10^9 Pa$. Hence, the sheer strain is

$$\gamma = \frac{\tau}{G} = \frac{625Pa}{25.5 \times 10^9 Pa} = 24.5 \times 10 - 9$$

The displacement in this case

$\Delta x = \gamma \times h = 24.5 \times 10^{-9} \times 0.1m = 2.45 \times 10^{-9}m$ or 2.45 nm.

What about fluid? In the case of fluid, we know that deformation will be continuous as long as the force is applied. To visualize this response, we illustrate the deformed body of fluid under a constant force at three different time instances as depicted in figure 10.3. Unlike the case of solid where deformation reaches a constant stage under the application of a constant force, here it keeps on increasing. If we assume that the rate of increase of this deformation is constant, hence one can state that for the case of fluid, the sheet stress is proportional to the rate of change of the shear strain, or

$$\tau \propto \frac{d\gamma}{dt} \tag{10.5}$$

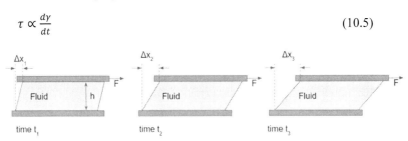

Figure 10.3: Deformation of a fluid body under constant shear stress as time progresses.

This relation in 10.5 happens to be true for most of the known liquids and all gasses. The proportionality can be written as an equality by introducing a proportionality constant, μ.

$$\tau = \mu \frac{d\gamma}{dt} \tag{10.6}$$

Where the constant μ is known as the fluid dynamic viscosity. From equation 10.6 one can deduce that the unit of viscosity is $Pa \cdot s$ or $N \cdot s/m^2$.

> *Viscosity is the resistance of the fluid to deformation at a given rate. It has units of $Pa \cdot s$ or $N \cdot s/m^2$*

The expression in equation 10.6 is also known as **Newton's law of viscosity**. Substances that follow the expression in equation 10.6 are called **Newtonian fluids**. Based on Newton's law of viscosity in equation 10.6 one could state that Newtonian fluid with high viscosity would experience less rate of change of the shear strain (less deformation rate) under a constant sheer stress. Viscosity of the fluid is also temperature dependent and as we could logically predict that viscosity of the fluid would decrease as temperature increases. Hence, table 10.2 shows typical values of dynamic viscosity for some fluids at room temperature.

Table 10.2: Dynamic viscosity at room temperature for different Newtonian fluids.

Fluid	μ in mPa.s $(10^{-3} \times Pa \cdot s)$	Fluid	μ in mPa.s $(10^{-3} \times Pa \cdot s)$
Water	0.89	Olive oil	56.2
Mercury	1.526	Honey	25.5
Milk	2		

Example 10.2:

A body of water at room temperature is placed between two plates that are separated by 1 mm such that it formed an almost rectangular shape of 5 mm width and 40 mm length. A constant force of 1 μN is constantly applied on the top plate. What is the displacement of the water after 1 seconds of the force application? What is the displacement if the fluid was replaced by olive oil?

Solution:

The shear stress is

$$\tau = \frac{F}{A} = \frac{10^{-6}N}{0.005m \times 0.04m} = 0.005Pa$$

The water viscosity at room temperature is $0.89 \times 10^{-3} \, Pa \cdot s$.

The rate of the shear strain is

$$\frac{d\gamma}{dt} = \frac{\tau}{\mu} = \frac{0.005Pa}{0.89 \times 10 - 3Pa \cdot s} = 5.617s^{-1}$$

The strain as a function of time is $\gamma = 5.617 \times t$

After one second, the displacement is then $\Delta x = 5.617 \times 1mm = 5.617mm$

For the case of olive oil, $\mu = 56.2 \times 10^{-3} \, Pa \cdot s$. The rate of the shear strain is

$$\frac{d\gamma}{dt} = \frac{\tau}{d\mu} = \frac{0.005 Pa}{56.2 \times 10^{-3} Pa \cdot s} = 0.089 s^{-1}$$

The strain as a function of time is $\gamma = 0.089 \times t$

After one second, $\gamma = \Delta x/h = 0.089 \times 1 = 0.089$

The displacement is then $\Delta x = 0.089 \times 1mm = 0.089mm = 89\mu m$.

10B. Fluid statics

In the previous section, we draw a clear line to distinguish between fluid
and solid based on their response to shear stress. The main distinction
between them is that solid deformations stop at a certain time and remain
constant as long as the shear is applied. However, for fluid this deformation
continues as long as the force is applied. The rate of deformation is constant
for the case of Newtonian fluids. The question here is can we find a way to
distinguish between gas and liquid in a similar way we did between solid
and fluid?

Density of the fluid

The main difference between these two phases of matter lies in the
separation between the atoms that form these fluids. Liquid atoms are in
close contact with each other. However, in contrast to solid, they could slide
over one another. Gasses on the other hand have large separation between
atoms. To keep gas in a certain volume one needs to place it in a container,
otherwise the molecules will disperse freely in space. For a fluid that is
contained in a certain volume V, the total mass of the fluid could be thought
of as the multiplication of the number of the atoms by the atomic mass.
Hence, denser fluid will have a larger number of atoms inside the volume
compared to low density matter. This results in a larger mass compared to
the low-density ones. That brings forward a quantity we call fluid density
that is defined by the mass of the fluid divided by the volume it occupies.

$$\rho = \frac{m}{V} \ kg \cdot m^{-3} \qquad\qquad (10.7)$$

Hence, one could easily build an argument that liquids are expected to have
much larger density compared to gas. This is actually the case when we
compare the density of water at room temperature, which is 997 kg/m^3, to
that of Helium gas at the same temperature under average atmospheric
pressure, which is 0.1607 kg/m^3. Like viscosity, density decreases with
temperature. We know from daily observations that increasing the water
temperature to its boiling point (100 $^\circ$C for water), it transfers to vapor
which has lower density than liquid water. This is mainly due to the increase
of the kinetic energy of the water molecules where gaps start to increase
between the molecules. There is an exception to this though around the
melting point (0 $^\circ$C for water) where the liquid is transferred into solid.
Between temperatures of 0 $^\circ$C to 4 $^\circ$C the density of water increases in spite
of the increase of temperature.

Constant volume

Another distinction between gas and liquid is the volume they occupy. For liquid, as per observations, when it is moved from one container to another, its volume remains constant. That is due to the presence of strong intermolecular force in liquid. The effect generates an interface (surface) between the liquid and the surrounding vapor/gas. At this interface, surface tension is observed where the liquid molecules have a stronger tendency to stack with each other (what is known as **cohesion**) rather than being attracted to the surrounding molecules (what is known as **adhesion**).

> **Cohesion** *is the property when molecules of the same kind tend to stick to each* *other.*
> **Adhesion** *is the property when molecules of different kidneys have strong attraction*

This is however not the case for gas. Gas molecules are largely separated, and the intermolecular force is weak. Hence, they tend to fill the volume of the container it occupies. If we move a certain mass of gas from a container of a small volume to a bigger one, we expect its density to decrease as it will spread over the new volume that it now occupies.

Compressibility

Compressibility is defined as the ability of the fluid (or solid) to change its volume under the application of pressure. When a force is normally applied to a surface, then pressure is simply the ratio between the force to the surface area.

$$P = \frac{F}{A} N \cdot m^{-2} \, orPa \tag{10.8}$$

This sounds very similar to shear stress. Shear stress is due a force that is applied parallel to the surface. Pressure is due to a force that is normal to the surface. This is visualized in figure 10.4 for the case of two containers of the same volume. One contains liquid while the other contains gas.

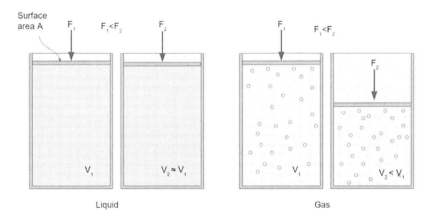

Figure 10.4: Compressibility of gas compared to liquid. Increasing the force reduces the gas volume however liquid almost experienced no change due to stronger intermolecular force.

When a force applied on the top surface is increased ($F_2 > F_1$) the liquid will experience minimal, almost negligible, compression (reduction of its volume). Gas on the other hand is compressed and its volume will reduce ($V_2 < V_1$) when the force increases. In terms of pressure, we could say that when a pressure $P_1 = F_1/A$ is applied then the gas reaches a volume V_1 when the pressure of the gas equals that of the outside ($P_{gas} = P_1$). When the pressure increases, $P_2 = F_2/A > P_1$, the gas volume is reduced to V_2 and its pressure increases to $P_{gas} = P_2$. The product of the pressure by the volume gives us a quantity that has units of energy.

$$P \left(\frac{N}{m^2} \right) \times V(m^3) = N \cdot m \tag{10.9}$$

The energy has units of Joule $J = N \cdot m$. Hence, we could argue that if the process of compressing the gas did not experience any loss of energy or mass, the produce of PV should be the same before compression and after compression. In other words

$$P_1 V_1 = P_2 V_2 \tag{10.10}$$

This is known as **Boyle's law**. We will be explaining this law in further detail in the next section. For now, we need to know that this law assumes that the temperature of the gas did not change during the compression. Another point to mention is that the gas here is assumed to be an ideal gas, where inter-particle interaction is neglected.

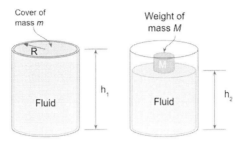

Figure 10.5: Cylindrical contain with a fluid that is compressed under the application of an external weight.

Example 10.3:

Consider two cylindrical containers of 5 cm radius and 15 cm of height. Both containers are filled with a fluid so that a circular cover of mass 1 kg is at the top of the container. The first fluid is water and the second is helium gas with pressure that is high enough to keep the cover at the top. If a mass of 2 kg is placed on top of the cover, what would be the height of the fluid in each container?

Solution for gas:

Originally the fluid has a pressure that equals the weight of the cover divided by the cover's area

$$P_1 = \frac{mg}{\pi R^2} = \frac{1kg \times 9.81m/s}{3.1416 \times (0.05m)^2} = 1249Pa$$

The original volume of the gas is

$$V_1 = h_1 \times \pi R^2 = 0.15m \times 3.1416 \times (0.05m)^2 = 0.0012m^3$$

The produce $P_1 V_1 = 1249Pa \times 0.0012m^3 = 1.471J$

The new force applied on the fluid when extra weight of mass M is applied is

$$F_2 = (m + M)g = (1kg + 5kg) \times 9.81m/s^2 = 58.7N$$

The pressure on the gas is

$$P_2 = \frac{F_2}{\pi R^2} = 7473.9 Pa$$

The new volume is obtained as follows

$$P_2 V_2 = P_1 V_2 \rightarrow V_2 = \frac{P_1 V_1}{P_2} = \frac{1.471 J}{7473.9 Pa} = 0.0002 m^3$$

The height is obtained as follows

$$V_2 = \pi R^2 h_2 \rightarrow h_2 = \frac{V_2}{\pi R^2} = \frac{0.0002 m^3}{3.1416 \times (0.05 m)^2} = 0.025 m$$

Alternative solution:

We know that $P_2 V_2 = P_1 V_1$. We also know that

$$P_1 = \frac{mg}{\pi R^2} \, and \, V_1 = h_1 \pi R^2 \, hence$$

$$P_1 V_1 = \frac{mg}{\pi R^2} \cdot h_1 \pi R^2 = mgh_1$$

That is the potential energy of the cover when it is placed at height h_1 (height of the fluid in the first case).

Similarly, we can find that $P_2 V_2 = (m + M)gh_2$. Hence, we can write that

$$mgh_1 = (m + M)gh_2 \rightarrow h_2 = \frac{mh_1}{m + M} = \frac{1kg \times 0.15m}{1kg + 5kg} = 0.025m$$

For liquid

$$V_1 \approx V_2 \, and \, hence \, h_1 = h_2 = 0.15m$$

The compressibility of fluid is typically represented by bulk modulus which is the ratio between the change in the applied pressure to the normalized change of the volume. Mathematically we write bulk modulus as

$$E_v = \frac{\Delta P}{\Delta V / V} \tag{10.11}$$

Bulk modulus has units of pressure or Pa. For instance, bulk modulus for water is $2.2\ Gpa$ or $2.2 \times 10^9\ Pa$. For the case of ideal gas with no change of temperature under pressure (isothermal), we know that $PV = constant$. Hence, we can use the difference operator to write that

$$\Delta PV + \Delta VP = 0 \tag{10.12}$$

We can arrange the equation above to become

$$\frac{\Delta P}{\Delta V/V} = -P \tag{10.13}$$

We know that when the pressure increases the volume decreases. Hence, when ΔP is positive then ΔV is negative. If we only assume the amplitude of change, then the negative sign at the right side can be omitted and the bulk modulus for gas becomes

$$E_v = P \tag{10.14}$$

Liquid pressure

In the previous section we talked about compressibility of gas. We stated that when increasing the applied pressure, the volume is reduced and hence the height of the cover in the cylinder example is reduced. What keeps the cover at the new height is that the pressure of the gas increases until a height is reached where the pressure inside matches that outside. In other words, the force applied on the cover area that pushes it down (gravity) equals the force pushing the cover up that is produced by multiplying the gas pressure inside by the cover area. This balance keeps the cover at the specific height. If this the case of gas, then what about liquid? How can we estimate the pressure inside the liquid?

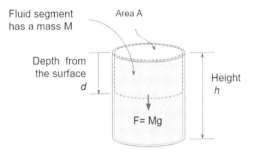

Figure 10.6: The body of liquid of depth d and area A has a mass M that generates a pressure P = Mg/A at that depth.

Consider a tank of height h that is filled to the top with a liquid. At a depth d from the surface, the body of the liquid will have a considerable weight Mg, where M is the mass of the liquid in that segment. We can calculate the liquid mass by multiplying the density by the volume the liquid occupies as in equation 10.7.

$$M = \rho A d \qquad\qquad (10.15)$$

If a thin sheet of area A is placed at depth d from the surface, then there will be a gravitational force of Mg that is exerted on the surface. This result in a pressure of

$$P = \frac{Mg}{A} = \rho g d \qquad\qquad (10.16)$$

The pressure inside the liquid increases by the depth.

Example 10.4a:

A submarine is submerged to a depth of 100 m beneath the water level. If the saltwater density is 1020 kg/m3, what is the pressure on the submarine body?

Solution:

The pressure from equation 10.16 is

$$P = \rho g d = 1020 kg/m^3 \times 9.81 m/s^2 \times 100m = 1000620 Pa \approx 1 MPa$$

We know that one bar ($\sim$ atmospheric pressure) is 100,000 Pa. Hence, the pressure on the submarine can be written as ·

$P = 1MPa = 10bar.$

Example 10.4b:

If the maximum water pressure human body can withstand is 6 bars, what is the maximum depth a diver can possibly reach?

Solution:

We know that $1bar \approx 100,000Pa$. The pressure is then 600,000 Pa.

From equation 10.16, the depth can be written as

$$P = \rho g d \rightarrow d = \frac{P}{\rho g} = \frac{600,000Pa}{1020\frac{kg}{m^3} \times 9.81\frac{m}{s^2}} \approx 60m$$

Surface tension

In chapter five we discussed the tension in the robe due to a weight that is attached to it. The gravity pulls the robe downwards and a tension force in the robe is acting oppositely. This tension was due to a pulling force caused by another weight hanging from the other side of a pulley. The presence of two opposite force on the two ends of the rope caused the tension inside it. In the case of the liquid molecules at the surface as in figure 10.7A, they exhibit cohesion force due to the surrounding molecules, which is larger than the adhesion with the surrounded air. Let us imagine a virtual robe formed by a line of molecules at the surface. At the microscopic scale we know each molecule on the surface experiences a pull from each side. Hence, there is a tension built inside the virtual robe. Unlike the cause in chapter five where the force causing the tension are localized at the two ends, here we have a distributed adhesion force along the robe. Hence, when dealing with surface tension, it would be more useful to define a force per unit length. We use the symbol σ for it and it has units of N/m.

This tension acts like the circus robe that a person could walk or bike on. Hence, a metallic needle or an insect could stay on the water surface. If we can imagine multiple of these virtual robes stretch on the liquid surface in almost all direction, then a surface is formed. That acts like a skin around the liquid volume causing droplets to form.

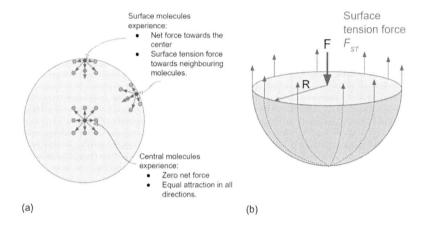

(a)   (b)

Figure 10.7: (a) Cohesion force between liquid molecules inside and at the surface of a water droplet. (b) Free body diagram of the forces acting on half a droplet.

Surface tension can be defined as surface force per unit length, and it usually has a symbol σ with units of N/m.

Water droplet

Let us examine the case of a water droplet of volume V. If we look at the molecules inside it, then the ones at the center would experience intermolecular (cohesion) forces that are equal in all directions. That gives a net zero force. However, the one at the surface would experience a net force that is pointing normal to the surface and towards the center of the droplet as shown in figure 10.7a. Surface molecules experience as well an equal pull from each side along the surface of the droplet due surface tension. This distributed tension across the surface tends to minimize the surface area around the volume of the liquid giving the spherical shape of the droplet. For a known volume of the droplet, the radius is

$$V = \frac{4}{3}\pi R^3 \rightarrow R = \left(\frac{3V}{4\pi}\right)^{1/3} \qquad (10.17)$$

In terms of density, we know that $\rho = \frac{m}{V}$ and hence, if we know the mass of the droplet, its

radius becomes

$$R = \left(\frac{3m}{4\pi\rho}\right)^{1/3} \qquad (10.18)$$

Example 10.5:

What is the radius of a water droplet of mass 10 grams in the atmosphere?

Answer:

Water density at room temperature is 997 kg/m³. Hence, the radius of the droplet is

$$R = \left(\frac{3 \times 0.01g}{4 \times 3.14 \times 997 kg/m^3}\right) = 0.013m = 1.3cm$$

If we consider the lower half of the droplet at in figure 10.7b, then we can draw few virtual lines (dashed lines). We know that the tension force right at the end of each line is pointing outwards due to the pull from the surrounding molecules. The total tension around the circumstance of the circle is

$$F_{ST} = 2\pi R\sigma \qquad (10.18a)$$

For the water droplet to sustain the shape it has, there should be a balance between the surface tension and the forces applied on the droplet, inner and external. This force is pointing upwards. It should hence balance an opposite force that is pressing downwards as in the figure. This force can be calculated from the difference between the pressure inside, P_i, and the droplet and outside it, P_e.

$$F = \pi R^2 (P_i - P_e) \qquad (10.18b)$$

Here, the force is the multiplication of the pressure difference by the circle area. At the state of equilibrium (when there are no more changes in the system) the two forces are balanced, $F_{ST} = F$.

$$\pi R^2 (P_i - P_e) = 2\pi R\sigma \tag{10.19a}$$

Or

$$P_i = P_e + \frac{2\sigma}{R} \tag{10.19b}$$

Hence the pressure inside the drop is greater than the surrounding. This is logical as otherwise the droplet would have collapsed inwards.

Example 10.6:

What is the pressure inside the water droplet in example 10.5 if the water droplet is formed in a pressure of 1 atmosphere?

Answer:

The external pressure is $Pe = 1\ atm = 101325\ Pa$.

The water tension at 25° is 0.072 Pa. Hence, the pressure inside the water droplet is

$$P_i = P_e + \frac{2\sigma}{R} = 101325 Pa + 2 \times \frac{0.072}{0.013} = 101327.6 Pa$$

Capillary action

One of the observed results of surface tension is the capillary effect. When a narrow tube of glass for instance is placed vertically inside a liquid, one would observe a rise or drop of the liquid level inside the tube. Let us examine water for instance. When a glass tube is inserted in the water, the adhesion force between the water molecules and the glass exceeds the cohesion force between the water molecules. Let's examine the inner circle of the tube that is at the water surface. The water molecules on the circle experience an upward pull by the tube inner wall. This force exceeds the surface tension. Hence, the surface of the water curves towards the tube walls in a convex shape. This pull keeps on moving the liquid surface upwards until it is balanced by the weight of the raised liquid. At this point the liquid height inside the tuber, h, remains constant as in figure 10.8a.

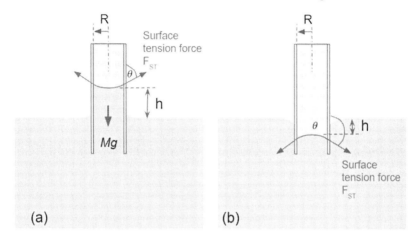

Figure 10.8: Capillary action when the contact angle is (a) less than 90° and (b) higher than 90°.

In the figure, the direction of the surface tension is pointing at an angle θ relative to the tube wall. This angle is referred to as the contact angle between the liquid and the tube material. At equilibrium the vertical component of the force

$$F = F_{ST} \times cos\theta = 2\pi R\sigma \times cos\theta \qquad (10.20)$$

has to equal the weight of the raised liquid,

$$F_g = Mg = \rho Vg = \rho(\pi R^2 h)g \qquad (10.21)$$

Equating the two forces, we obtain

$$2\pi R\sigma \times cos\theta = \rho(\pi R^2 h)g \qquad (10.22)$$

Hence, the height of the water column is

$$h = \frac{2\sigma cos\theta}{\rho R g} \qquad (10.23)$$

In some cases, such as mercury, the contact angle is larger than 90°. This is because the cohesion force between the liquid molecules is higher than the adhesion force between the liquid and solid molecules forming the tube surface. That results in a negative sign in the cosine term. The liquid level inside the tube then tends to move down the tube instead of rising as depicted in figure 10.8b.

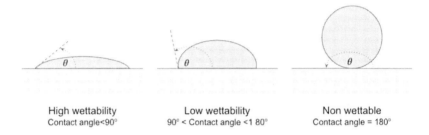

High wettability   Low wettability   Non wettable
Contact angle<90°   90° < Contact angle <1 80°   Contact angle = 180°

Figure 10.9: Capillary action when the contact angle is (a) less than 90° and (b) higher than 90°.

It is worth noting that the contact angle is typically used to determine the wettability of a solid surface. Zero contact angle indicates perfect wetting. Contact angles less than 90° define a highly wettable surface. Angles larger than 90° and lower than 180° define low wettability. Liquids with 180° contact angle are non-wettable. These contact angles are depicted in figure 10.9. In the case of water, surfaces that are wettable are referred to as **hydrophilic**. Surfaces that are non-wettable are often called **hydrophobic**.

Example 10.7:

Three glass tubes were placed in a water tank. The tubes have radii of 0.5 mm, 1 mm and 3 mm respectively. Estimate the height of the liquid inside the tube compared to the water surface outside if the contact angle between water and glass when surrounded by air at room temperature is 45°?

Solution:

From equation 10.23, the height is $h = 2\sigma cos\theta/\rho R g$

The contact angle is $\theta = 45^o$.

Water density at room temperature is 997 Kg/m3.

Surface tension between water-air at room temperature is $\sigma = 0.072$ N/m

The heights for the three tubes are:

$R = 0.0005m \Rightarrow$

$$h = 2 \times 0.072 \times cos(45^o)/(997 \times 0.0005 \times 9.81) = 0.0208m$$
$$= 2.08cm$$

$R = 0.001m \Rightarrow$

$$h = 2 \times 0.072 \times cos(45^o)/(997 \times 0.001 \times 9.81) = 0.0104], m$$
$$= 1.04cm$$

$R = 0.003m \Rightarrow$

$$h = 2 \times 0.072 \times cos(45^o)/(997 \times 0.003 \times 9.81) = 0.00347m$$
$$= 3.47mm$$

Buoyancy

When an object of mass M is placed on the liquid surface, the gravitational force tends to push the object downwards. That causes the object to displace the fluid under it starting to submerge as in figure 10.10. However, we know from equation 10.16 that the liquid pressure increases with the depth. Hence, the object experiences higher pressure at the bottom compared to the top. This difference in pressure results in a force that pushes the object upwards and opposes its submergence. This force is known as Buoyancy. This what causes objects within certain limits to float on the water. The buoyancy force is

$$F_B = \rho V_{Disp} g \tag{10.24}$$

Here V_{Disp} is the displaced volume of the liquid due an object of mass m. When the object floats, this force balances the gravitational force, $\rho V_{Disp} g = mg$. Hence, the displaced volume is

$$V_{Disp} = m/\rho \tag{10.25}$$

If an object with flat sides (such as the cylinder in figure 4.10b) is placed on water, then the displaced volume is $V_{Disp} = Ad$, where d is the depth of the submerged section of the object. It can be estimated as

$$d = \frac{m}{\rho A}$$ (10.26)

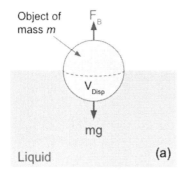 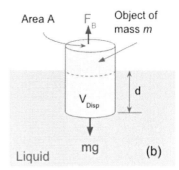

Figure 10.10: An object of mass m floating on a water when the buoyancy force balances the gravitational force. (a) Spherical object of mass m displaces water of volume V_{Disp}. (b) An object with flat sides of area A has a submergence depth d.

Example 10.8:

Two Cylinders of equal radius, 10 cm, and height, 10 cm, were placed on the water. One cylinder is metallic while the other is made of wood. The metallic cylinder has a mass of 2 Kg, and the wooden one has a mass of 500 grams.

(a) Would both cylinders float?

(b) What are the depths of the submerged segments of the floating cylinder(s)?

(c) If the metallic cylinder dimensions remain the same, what would be the minimum mass needed for the cylinder to submerge underwater?

(d) If the metallic cylinder kept the same mass and height, what would be the minimum radius such that the object still floats?

Solution:

Water density at room temperature is 997 Kg/m³.

The volume of each cylinder is $V = \pi R^2 h = \pi \times 0.1^2 \times 0.1 = 0.00314 m^3$

(a) In the case of the metallic cylinder, mass=2 kg, the displaced volume is

$V_{Disp} = m/\rho = 2/997 = 0.002 m^3$

This volume is smaller than the cylinder volume, hence it will float.

In the case of a wooden cylinder

$V_{Disp} = m/\rho = 0.5/997 = 0.0005 m^3$

This volume is smaller than the cylinder volume, hence it will float as well.

(b) For the metallic cylinder the submerged segment is

$d = V_{Disp}/A = 0.002/(\pi \times 0.12) = 0.063 m = 6.3 cm$

For the wooden cylinder he submerged segment is

$d = V_{Disp}/A = 0.0005/(\pi \times 0.12) = 0.016 m = 1.6 cm$

(c) For the cylinder to just submerge totally, the volume of the displaced liquid should equal that of the cylinder, $V = V_{Disp}$. Hence, the mass of the object should be

$V_{Disp} = m/\rho = V \rightarrow m = \rho V = 997 \times 0.00314 = 3.13 kg$

(d) For the object to just float then $V = V_{Disp} = \pi R^2 d$. If the object kept the same height, then d=10 cm. Using equation 10.26

$d = m/\rho A \rightarrow A = m/\rho d = 2/997 \times 0.1 = 0.02 m^2$

The radius is then

$R = 0.02/\pi = 0.08 m = 8 cm$

10C. Fluid dynamics

In the previous sections, the fluid in all cases was assumed to be at rest and the system temperature does not change. In that case the fluid energy at any state is only in the form of potential energy. In this section, the fluid is assumed to move, and its speed is assumed to be uniform across the area it passes through.

Fluid flow rate

Let us assume that an incompressible fluid such as liquid moves with a constant speed u in a pipe that has a cross-sectional area A as shown in figure 10.11a. During time interval dt a volume dV will pass through the shaded area in the figure. This volume is the multiplication of the area by the distance the liquid covers during the time dt.

$$dV = (u \cdot dt)A \tag{10.27}$$

In other words, the amount of volume crossing the area A per unit time is

$$dV/dt = u\,A \tag{10.28}$$

This ratio in equation 4.28 is commonly known as the volume flow rate or Q and it has units of m³/s, $Q = dV/dt = uA$. In terms of mass, we know that for incompressible fluid with constant density ρ, the mass is the multiplication of the density by the volume. Hence, the rate of change of the mass is

$$\frac{dm}{dt} = \frac{d(\rho V)}{dt} = \rho \frac{dV}{dt} = \rho u A \tag{10.29}$$

This is known as the mass flow rate or ṁ and it has units of kg/s. Hence, we can write equation 10.29 as $\dot{m} = \frac{dm}{dt} = \rho u A$.

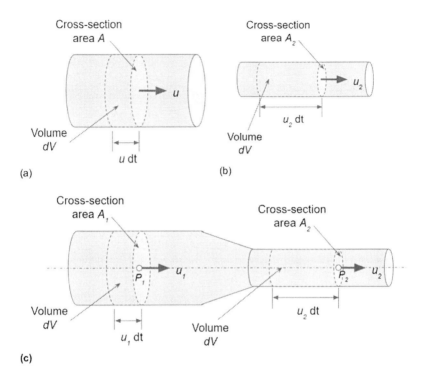

Figure 10.11: (a) A fluid flowing through a pipe of cross section A with a uniform speed, u, results in a volume $dV = A\,u\,dt$ to pass the cross-section within time dt. (b) When the same volume passes through a smaller area A_2 during time dt causes a higher flow speed u_2. (c) Flowing from a large cross section area to a small one increases the speed of flow.

Now, let us assume that same fluid passes through a smaller pipe with cross section $A_2 < A_1$. Here we are referring to the cross-section area of the first segment as A_1. The flow rate inside the small pipe is $Q_2 = \frac{dV}{dt} = u_2 A_2$ where u_2 is the speed of the fluid inside the pipe as shown in figure 10.11b. The two pipes are connected together (figure 10.11c) such that the flow rate in both sections are the same. During a time, interval dt, the same volume dV/dt flows in both sections.

$$\frac{dV}{dt} = u_1 A_1 dt = u_2 A_2 dt \rightarrow u_1 A_1 = u_2 A_2 \qquad (10.30)$$

The ratio between the two speeds is then

$$\frac{u_2}{u_1} = \frac{A_1}{A_2} \qquad (10.31)$$

Hence, when the cross-section area is reduced, then the speed of the fluid increases when considering that the fluid is incompressible. However, one could state an apparent contradiction: the kinetic energy of the fluid (a fluid droplet of mass δm and volume δV for instance) in the second segment has increased.

$$\Delta KE = KE_2 - KE_1 = \frac{1}{2}\delta m u_2^2 - \frac{1}{2}\delta m u_1^2 = \frac{1}{2}\delta m (u_2^2 - u_1^2) \qquad (10.32)$$

According to the law of conservation of energy, the total energy in the first segment should equal that of the second. If there is no change in the potential energy (both droplets of water are in the same horizontal plane), then what causes this increase in the kinetic energy? To answer this question, we need to understand the different types of energy in a fluidic system.

Fluid energy

The apparent contradiction presents arises becase we looked at the droplet as if it was a solid object moving in air under only the influence of gravity. In our case of study however the considered volume moves inside the fluid where other factor such as the fluid microscopic properties and pressure cannot be neglected. For example, the points p_1 and p_2 in figure 10.11, although they have the same gravitational potential energy, the liquid pressure at each point is different. You can think of it as the two points are at different depth relative the liquid surface (tube wall). That brings forward another type of potential energy that one needs to consider, which is pressure potential energy.

Potential energy

Potential energy of a body of fluid can be in two forms: gravitational potential energy and one due to pressure applied on the fluid. For a fluid of mass m located at a height z from the ground, it has a gravitational potential energy of

$$PE_g = mgz \qquad (10.33)$$

When dealing with fluids, it might be more practical to define the potential energy per unit mass.

$$pe_g = \frac{PE_g}{m} = gh \qquad\qquad (10.34a)$$

Another normalization is when considering the potential energy per unit volume.

$$pe_{g,v} = \frac{PE}{V} = \frac{m}{V}gz = \rho gz \qquad\qquad (10.34b)$$

Here, $\rho = m/V$ is the density of the fluid as defined earlier.

Another type of potential energy is that due to a pressure P on a point located at a depth d from the fluid surface. For a point at depth d from the surface we can define the potential energy as the work done, W, to move this point to that depth against the applied force, F, or $W = F \cdot d$. We know that pressure, P, is force per area and hence we can write force in terms of pressure as $F = P \cdot A$. Hence, the potential energy due to pressure equals the work

$$PE_p = W = P \cdot Ad = PV \qquad\qquad (10.35)$$

We know from equation 10.15 that the fluid pressure is $P = \rho gd$. Hence, we can write the potential energy as $PE_p = (\rho gd)V = (\rho V)gd$. We know that $\rho V = m$, where m is the mass of the liquid above above the point.

$$PE_p = mgd \qquad\qquad (10.36)$$

The potential energy due to pressure per unit mass is

$$pe_p = \frac{PE_p}{m} = gd \qquad\qquad (10.37a)$$

The potential energy per unit volume can be obtained from equation 10.35 as

$$pe_{p,v} = \frac{PE_p}{V} = P \qquad\qquad (10.37b)$$

Kinetic energy

When a fluid moves, it gains a kinetic energy as a whole that depends on its speed. Kinetic energy of a moving object is defined as $KE = \frac{1}{2}mu^2$, where m is the mass of a body of fluid (for example a droplet of water) that moves, and u is its speed. The kinetic energy per unit mass can be written as

$$ke = \frac{KE}{m} = \frac{1}{2}u^2 \tag{10.38}$$

In terms of volume, we can define the kinetic energy per unit volume as

$$ke_V = \frac{KE}{V} = \frac{1}{2}\rho u^2 \tag{10.39}$$

Where $\rho = m/V$ is the density of the fluid.

Internal energy

At the microscopic level, several types of potential and kinetic energies exist. For example, the motion of the fluid molecules represents the thermal energy and is indicated by the temperature in Kelvin as it will be explained in the next chapter. Chemical and electrical energies are examples of the internal potential energies that depend on the chemical and electrical properties of the fluid. These energies at the microscopic level are referred to as the internal energy, U_i. If we neglect any changes in the internal potential energy of the fluid, the change of U_i can mainly be attributed to the change in fluid temperature. Hence, if we assume that there is no change in the temperature then one can assume that the internal energy U_i of the fluid remains constant at all different points inside the fluid, which is the case of this chapter.

Bernoulli equation

Now that we know the different types of kinetic and potential energies in a fluidic system, we can examine the total energy of the system in figure 10.11c at two points: P_1 inside the first segment and P_2 inside the second segment. The total energy at each point is

$$E_1 = U_{i,1} + PE_{g,1} + PE_{p,1} + KE_1 \tag{10.40a}$$

$$E_2 = U_{i,2} + PE_{g,2} + PE_{p,2} + KE_2 \tag{10.40b}$$

The law of conservation of energy states that $E_1 = E_2$ as we do not assume any external work is applied on the system.

$$U_{i,1} + PE_{g,1} + PE_{p,1} + KE_1 = U_{i,2} + PE_{g,2} + PE_{p,2} + KE_2 \tag{10.41}$$

As mentioned earlier, in this chapter the fluid temperature is assumed to be constant. Hence,

The internal energy, U_i, remains constant as well ($U_{i,1} = U_{i,2}$). That reduces the equation above to

$$PE_{g,1} + PE_{p,1} + KE_1 = PE_{g,2} + PE_{p,2} + KE_2 \qquad (10.42a)$$

Or

$$mgz_1 + P_1 V + \frac{1}{2}mu_1^2 = mgz_2 + P_2 V + \frac{1}{2}mu_2^2 \qquad (10.42b)$$

In equation 10.42b, we are comparing two bodies of the fluid that have the same volume and mass. In other words, the density does not change as the fluid is incompressible.

Dividing both sides of the equation by the volume **V**

$$\rho g z_1 + P_1 + \frac{1}{2}\rho u_1^2 = \rho g z_2 + P_2 + \frac{1}{2}\rho u_2^2 \qquad (10.43)$$

Equation 4.43 is known as the **Bernoulli equation** for incompressible fluid.

In the example in figure 10.11c, both points in comparison are at the same height (at the same horizontal plane) then $z_1 = z_2$. That reduces equation 10.43 to

$$P_1 + \frac{1}{2}\rho u_1^2 = P_2 + \frac{1}{2}\rho u_2^2 \qquad (10.44)$$

Equation 10.44 clearly answers the apparent contradiction we discussed earlier. The increase in the kinetic energy of the fluid comes from the change of the pressure in both segments. We can write the change of pressure as

$$\Delta P = P_2 - P_1 = \frac{1}{2}\rho(u_1^2 - u_2^2) \qquad (10.45)$$

For the example in figure 10.11c, we know that the ratio between the speeds is $\frac{u_2}{u_1} = \frac{A_1}{A2}$. Hence, it is clear than $u_2 > u_1$ as $A_1 > A_2$. That results in a negative quantity for the quantity between parentheses. The pressure on the fluid body is then reduced when the speed is increased, $p2 < p1$. This is known as Bernoulli's principle.

> **Bernoulli equation:** equates the total energy at two different points inside the fluid.
>
> **Bernoulli principle:** When the speed of the fluid increase, then its pressure decreases.

Example 10.9:

Consider the system in figure 10.12. The small tube end is open to the air. The other side is connected to a pump that provides a pressure P_1 such that a constant volume flow of $Q= 1\,liter/s$ is sustained. The bigger tube has a diameter of 4 cm while the smaller one has a diameter of 1.5 cm.

a) What are the fluid speeds in both tubes?

(b) What is the pressure the pump needs to provide in order to sustain the desired flow rate?

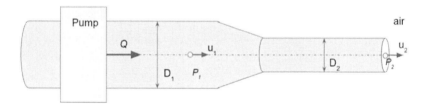

Figure 10.12: A Pump is applying a pressure p on the body of water at one end in order to sustain a constant volume flow rate Q. The other end of the pipe is open to the atmosphere.

Solution:

The cross-section areas of both sides are

$$A_1 = \pi R_1^2 = \pi (D/2)^2 = 3.14 \times (0.04/2)^2 = 0.0012 m^2$$

$$A_2 = \pi R_2^2 = \pi (D_2/2)^2 = 3.14 \times (0.015/2)^2 = 0.00018 m^2$$

The volume flow rate is $Q = 1\ liter/s = 1000\ cm^3/s = 0.001\ m^3/s$

(a) $Q = A_1 u_1 \rightarrow u_1 = Q/A_1 = 0.001/0.0012 = 0.833 m/s$

$Q = A_2 u_2 \rightarrow u_2 = Q/A_2 = 0.001/0.00018 = 5.5 m/s$

(b) From equation 10.45 the difference in the pressure

$$\Delta P = P_1 - P_2 = \frac{1}{2}\rho(u_2^2 - u_1^2)$$

Water density is $\rho = 997 \ kg/m3$

$$\Delta P = P_1 - P_2 = \frac{1}{2}\rho(u_2^2 - u_1^2)$$

$$\Delta P = P_1 - P_2 = \frac{1}{2} \times 997 \times (5.5^2 - 0.833^2) = 14733 Pa$$

$$P_1 = P_2 + 14733 Pa$$

Point P2 is exposed to air, hence the pressure is that of the atmosphere, 101,325 Pa. Hence

$$P_1 = 101325 + 14733 = 116058 Pa$$

The pump needs to provide 116058 Pa in order to achieve the desired flow rate.

Example 10.10

What pressure does a pump need to produce in order to generate a water flow rate of 2 liters per second in a 2 cm diameter tube from the ground level to a height of 3 meters as shown in figure 10.13? The other end of the tube is open to the atmosphere.

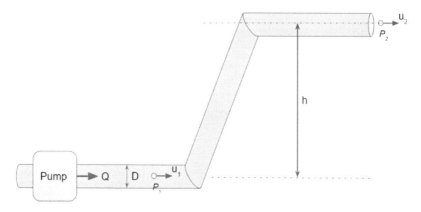

Figure 10.13: A Pump is pushing the liquid to flow to an elevation h.

Solution:

Using Bernoulli equation between the points P_1 and P_2

$$\rho g z_1 + P_1 + \frac{1}{2}\rho u_1^2 = \rho g z_2 + P_2 + \frac{1}{2}\rho u_2^2$$

The water flow is $Q = 2\ liter/s = 2000\ cm^3/s = 0.002\ m^3/s$

The pipe area is the same, and assuming a constant flow rate, then the speed of the liquid does not change, $u_1 = u_2$.

The second end is open to atmosphere, then $P_2 = 1\ atm = 101325\ Pa$

Water density at room temperature is $\rho = 997\ kg/m^3$

Bernoulli equation is simplified to

$$\rho g z_1 + P_1 = \rho g z_2 + P_2$$

$$P_1 = P_2 + \rho g(z_2 - z_1) = P_2 + \rho g h$$

$$= 101325 + 997 \times 9.81 \times 3 = 130666.71\ Pa$$

Flow from a fluid tank

Let us consider the case when a tank of fluid is filled to the top and a hole is placed at a depth d from the surface. So, we can apply Bernoulli's equation between two points: P_1 at the surface and P_2 right outside the hole.

If we assume that both points are exposed to the atmosphere then Bernoulli equation can be simplified as

$$\rho g z_1 + \frac{1}{2}\rho u_1^2 = \rho g z_2 + \frac{1}{2}\rho u_2^2 \tag{10.46}$$

If we neglect the speed of liquid at the surface, $u_1 \approx 0$, then equation 10.46 is reduced to

$$g z_1 = g z_2 + \frac{1}{2} u_2^2 \tag{10.47}$$

The speed of the fluid flowing through the hole is then

$$u_2 = \sqrt{2g(z_2 - z_1)} = \sqrt{2gd} \tag{10.48}$$

A question remains here, "how far would a flow of water coming out of the hole travels in the horizontal direction when it reaches the ground?" This can be answered using Newton's second law of motion where we assume a body of liquid of mass m starts at an initial speed of u_2 pointing horizontally, along the x axis.

The initial velocity is then, $\overline{u}_o = u_2 \hat{x}$. Gravity is on the other hand pulling the water body down with an acceleration of g pointing in the negative y direction. The initial displacement is $\overline{y}_o = h\hat{y}$. Hence,

$$\overline{a} = 0\hat{x} - g\hat{y} \tag{10.49}$$

Velocity is the integration of acceleration

$$\overline{u} = \overline{u}_o + \int \overline{a}\,dt = u_2 \hat{x} - gt\hat{y} \tag{10.50}$$

Displacement is the integration of velocity

$$\overline{r} = \overline{r}_o + \int \overline{u}\,dt = u_2 t\hat{x} + \left(h - \frac{1}{2}gt^2\right)\hat{y} \tag{10.51}$$

When the water reaches the ground the y values should be zero, hence

$$t_{max} = \sqrt{2h/g} \tag{10.52}$$

The maximum horizontal distance the water travels when it hits the ground is

$$x_{max} = u_2 t_{max} = u_2\sqrt{2h/g} \tag{10.53}$$

Example 10.11:

For the tank in example 10.11, if the tank is 1 meter high and it is placed on a table that is 60 cm above the ground. A hole is made right near the bottom of the tank. What is the speed of the water flowing from the tank and what distance would it cover when it reaches the ground?

Solution:

Water has density ρ = 997 kg/m^3.

Depth of the hole, d = height of the tank = 1m.

Height from the hole to the ground, h = height of the table = 0.6 m.

From equation 10.48 the speed is

$$u_2 = \sqrt{2gd} = \sqrt{2 \times 9.81 \times 1} = 4.43 m/s$$

From equation 10.53 the distance is

$$x_{max} = u_2\sqrt{2h/g} = 4.61 \times \sqrt{2 \times 0.6/9.81} = 1.8m$$

10D. Summary

- Solid stop getting deformed after a short time of force application.
- Fluid is a substance that deforms continuously under the application of external shear stress.
  - Shear force: force per unit area of the surface
    - $\tau = \frac{F}{A} N/m^2$
  - Shear strain: Lateral displacement/thickness
    - $\gamma = \frac{\Delta x}{h}$
- For solid
  - Strain is proportional to shear force
    - $\tau = G\gamma$
    - G is sheer modulus of the solid
- For liquid
  - Rate of change of strain is proportional to shear force
    - $\tau = \mu \frac{d\gamma}{dt}$
      - μ is known as the fluid dynamic viscosity
      - Viscosity is $Pa \cdot s$ or $N \cdot s/m^2$
  - Fluid density
    - Mass of the fluid divided by the volume
      - $\rho = \frac{m}{V} kg \cdot m^{-3}$
- Liquid
  - Cohesion is the property when molecules of the same kind tend to stick to each other.
  - Adhesion is the property when molecules of different kidneys have strong attraction
- Compressibility
  - Liquid volume remains constant due strong intermolecular force.
  - Gas volume changes with the container due to weak intermolecular force.
    - $PV = constnat$
    - Boyle's law
      - $P_1 V_1 = P_2 V_2$
    - Bulk modulus
      - $E_v = \frac{\Delta P}{\Delta V / V}$

- Liquid pressure
  - $P = \rho g d$
    - Depth of the surface: d
    - Volume density: ρ
- Surface tension
  - σ : Surface force per unit length, units of N/m
  - Water droplet
    - Radius
      - $R = \left(\frac{3m}{4\pi\rho}\right)^{1/3}$
      - Mass of the droplet: m
    - Tension force around the droplet
      - $F_{ST} = 2\pi R\sigma$
    - Liquid pressure inside the droplet
      - $P_i = P_e + \frac{2\sigma}{R}$
  - Capillary action
    - Height of water in the column is
      - $h = \frac{2\sigma \cos\theta}{\rho R g}$
    - Contact angle: θ
      - High wettability: $\theta < 90^o$
        - Hydrophilic
      - Low wettability: $90^o < \theta < 180^o$
        - Hydrophobic
      - Non wettable: $\theta = 180^o$
- Buoyancy
  - Displacement volume:
    - $V_{Disp} = m/\rho$
  - Displacement height:
    - $d = \frac{m}{\rho A}$
- Fluid dynamics
  - Volume flow rate
    - $Q = dV/dt = uA$ m^3/s
      - A : Tube area
      - u : Fluid speed
  - Flow between two tubes
    - $\frac{u_2}{u_1} = \frac{A_1}{A_2}$

- Fluid energy
  - Potential energy
    - Gravitational potential energy
      - Per unit mass: $pe_g = gh$
      - Per unit volume: $pe_{g,V} = \rho g h$
    - Pressure potential energy
      - Per unit mass: $pe_p = gd$
      - Per unit volume: $pe_{g,V} = P$
  - Kinetic energy
    - Per unit mass: $ke = \frac{1}{2}u^2$
    - Per unit volume: $ke_V = \frac{1}{2}\rho u^2$
  - Internal energy
    - The change of U_i can mainly be attributed to the change in fluid temperature.
- Bernoulli equation
  - $PE_{g,1} + PE_{p,1} + KE_1 = PE_{g,2} + PE_{p,2} + KE_2$
  - Per unit volume
    - $\rho g z_1 + P_1 + \frac{1}{2}\rho u_1^2 = \rho g z_2 + P_2 + \frac{1}{2}\rho u_2^2$

10E. Review questions

1. A cylinder of 10 cm radius and 20 cm height with a movable lid contains a gas which at equilibrium when the surrounding is at atmospheric pressure the lid is stabilized at a height of 15 cm measured from the bottom of the cylinder. A weight was added on top of the lid that causes the lid to move 2 cm downwards. Calculate the placed weight knowing that during this process the temperature of the gas inside did not change. We can neglect the weight of the cover lid.

2. If the gas inside the cylinder was replaced by water, and the same weight as in question 1 was placed on the top, how far downwards would the lid move?

3. If we remove the lid from the system in question 2 and the mass was assumed to be a small cylinder of 4 cm radius and 5 cm height, then would that weight sink or float. If it floats, how deep would that weight submerge?

4. If the mass of the small cylinder in equation 3 was chosen so it submerges inside the water and sinks to the bottom, then calculate the liquid pressure that the top of the cylinder experiences?

5. The table at the bottom shows some values of the contact angle of different surfaces to the water/air interface.

    a. Based on the table list which surface has high or low wettability.
    b. If three tubes of identical diameter of 1 cm are made of glass, copper and Teflon are placed in water, what would be the water height inside each tuber relative to the water surface?

Surface	Contact angle (°)
Glass	30
Porcelain	45
Aluminum foil	74
Copper	88
Stainless steel	75
Candle	100
Teflon	139

6. Two water tubes of different radii are connected horizontally. The bigger tube has a 5 cm radius while the smaller one has a radius of 1 cm. If the tubes are connected using a tapered section and the water was flowing at a rate of 4 liters per second though the bigger tuber, what is the flow rate of the water coming out of the smaller tube? What is the speed of the fluid going through both tubes?

7. If the small tube is bent vertically to carry the water to the second floor that is 3 meters high. What is the speed of the water coming out in this case?

8. A water tank is made of cylindrical shape of 50 cm radius and 150 cm height. The water completely fills the tank. A water tube of 1 cm diameter is connected horizontally right at the bottom of the tank. Calculate the flow rate of the water going into the tube knowing that the other end of the tube is open to the atmosphere.

9. If the tank was placed at a height of 1 meter off the ground, how far away from the tank would the water reach the ground?

10. A pipe with a diameter of 5 cm has water flowing through it at a velocity of 10 m/s. If the pressure at point A is 2 atm and the elevation difference between point A and point B is 5 m, what is the pressure at point B?

11. A house is experiencing a category one hurricane of wind speed 120 km/h. If the pressure inside the house is at atmospheric pressure and the air density is 1.2 kg/m^3, then what is the lifting force a roof of an area of 4 m^2 experiences?

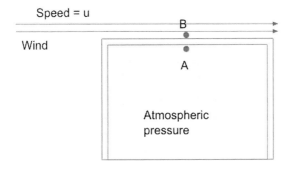

12. An airplane wing has a surface area of 20 m². If the air density is 1.2 kg/m³ and the air velocity over the wing is 80 m/s, what is the speed under the wing that is needed to start lifting the airplane if it has a mass of 800 kg.

13. A hydraulic system consists of a pipe with a diameter of 10 cm and a pump that delivers 100 L/min of oil at a pressure of 1000 kPa. What is the velocity of the oil in the pipe?

14. Water flows through a horizontal pipe that has a diameter of 2 cm. If the pressure at point A is 200 kPa and the pressure at point B is 150 kPa, what is the velocity of the water in the pipe?

CHAPTER ELEVEN

INTRODUCTION TO THERMODYNAMICS

11A. Introduction

Thermodynamics deals with the relationships between heat, energy, and work. We have covered the topic of energy and energy conversion in the previous book, where two forms of energy were considered: potential and kinetic energies. In general energy is the ability to do work. The potential energy of an apple held on the tree allowed it to fall down (do work) when it was detached. When dealing with fluids as in the previous unit, we assumed no temperature change and hence, we were concerned with the potential and kinetic energies of the fluid at different points. Applying the law of conservation of energy between two different points resulted in Bernoulli equations.

In that unit we introduced another form of energy, namely the **internal energy**. It deals with the microscopic state of the particles (atoms/molecules) that make up a system. It includes both the kinetic energy of the particles (related to their motion) and the potential energy of the particles (related to their position, configuration or chemical reactions). The kinetic energy part of the internal energy is commonly referred to as the **thermal energy of the system**. The internal energy as well includes other forms of energy such as: nuclear energy, electrical energy, chemical energy and magnetic energy.

Thermal energy

Thermal energy is a form of energy that is related to the motion of particles within a substance or system. This kinetic energy is transferred between the objects or systems in a form of heat. In other words, heat is the amount of kinetic energy transferred between objects or systems which have different temperatures, where temperature is a measure of the average kinetic energy of the particles in a system. It is often measured in degrees Celsius (°C) or Kelvin (K). The Kelvin scale is based on absolute zero, which is the temperature at which all particles in a system have zero kinetic energy.

Table 5.1. Definitions of thermodynamic quantities

Internal energy	Kinetic and potential energy of the substance at the microscopic scale.
Thermal energy	Kinetic energy of the particles forming the substance.
Temperature	The measure of the average kinetic energy of the particles.
Heat	The amount of the particles' kinetic energy that transfers from systems of different temperatures.
Absolute zero	The temperature at which all particles have zero kinetic energy.

Heat

Heat is the transfer of energy between two objects due to a temperature difference. When heat flows from a hotter object to a cooler object, the hotter object loses energy and the cooler object gains energy. Heat can be transferred by:

- Conduction (direct contact between objects)
- Convection (movement of a fluid)
- Radiation (emission of electromagnetic waves).

To better understand this, consider what happens when you boil a metallic pot filled with water on fire or electrical stove as in figure 11.1. The fire heats the lower part of the metallic pot. The heat is transferred to the walls of the pot and the handle by direct contact. This is called conduction. Inside the pot the fluid near to the bottom gets hotter and hence its molecule's kinetic energy is increasing making them move to the upper layers of the fluid. This motion causes the molecules in the upper layer to start moving as well. When moving down they gain more thermal energy and their temperature rises. This means of transferring heat through the motion of molecules in a fluid is referred to as convection. Finally, as we discussed in black body radiation, thermal sources such as fire emits electromagnetic radiation. This radiation carries an energy and hence heat can be transferred through this emission. This form of heat transfer is referred to as radiation.

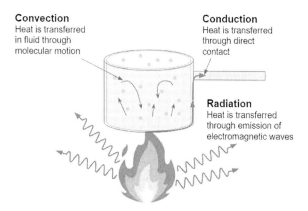

Figure 11.1. Heat transfer when heating liquid in a pot in three forms: conduction, convection and radiation.

Thermodynamic systems

In the example above we can consider a system that is composed of a pot with metallic walls and a loose cover. The pot contains a liquid substance, and the walls allow heat to transfer in and outside the system. The loose led allows work to be performed inside the system. The fact that the lid is not covered, allows the liquid vapor to move outside the system and hence in this case we considered this system as an open system.

If the lid is made tight enough that it does not allow the fluid or vapor to leave the pot, the system is then referred to as a closed system. Finally, if the enclosed liquid in the pot was made such that it does not have any interaction with the environment, then the system would be referred to as isolated. Figure 11.2 illustrates graphically the different thermodynamic systems.

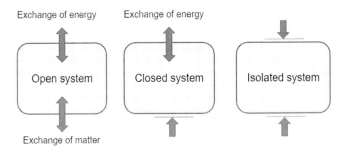

Figure 11.2. Different types of thermodynamic systems.

Another point here is the nature of the walls that enclose the material in the thermodynamic system. In the example of the cooking pot, having metallic walls allows conductive heat to transfer into and out of the pot. In this case the wall is diathermal. If the pot was made of a thermally insulating material that does not allow conductive heat transfer, the wall would become adiabatic. From the point of view of motion, if the wall is attached and not movable, then it is referred to as a fixed wall, otherwise it is called movable. For the case of the loose cover, it allows the material inside the pot to interact with the environment, e.g., evaporate and leave the system. In this case this wall is referred to as permeable, otherwise it is called impermeable. These different types of walls are depicted in figure 11.3.

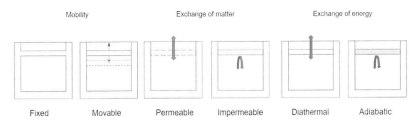

Figure 11.3. Types of walls in a thermodynamic system.

Table 11.2. Definitions for thermodynamic systems and walls

Open system	Allows transfer for both energy and matter.
Closed system	Allows only energy transfer.
Isolated system	Does not allow energy or matter transfer.
Fixed wall	Does not move, the opposite is movable
Permeable wall	Allows matter transfer, the opposite is impermeable.
Diathermal wall	Allows energy transfer, the opposite is adiabatic.

Thermodynamic processes

At any point of time, a thermodynamic system can be defined by a set of state variables: temperature, T, pressure, P, and volume, V. When a work is applied to the system or an energy is added, all or some of the parameters change from the initial state. These processes can be categorized based on the nature of the processes as in table 11.3.

Table 11.3. Thermodynamic processes based on their nature of work or energy applied.

Mechanical process	Mechanical work is applied on the system. For instance, it could reduce the volume, increase the pressure and possibly temperature.
Thermal process	The application of thermal energy. For instance, it can increase temperature and pressure. It could as well cause mechanical movements and hence changes volume.
Chemical process	It refers specifically to matter being exchanged between a system and its environment. This is not to be confused with chemical reactions among different substances.

The thermodynamic processes could as well be defined based on how they affect the state variables and energy transfer.

Table 11.4. Thermodynamic process based on the effect on the state variable.

Isothermal process	The temperature does not change during the process.
Adiabatic process	No thermal energy is allowed to flow in or out of the system.
Isobaric process	The pressure remains constant throughout the process.
Isochoric process	The volume remains constant throughout the process.

11B. The first law of thermodynamics

The first law of thermodynamics is the principle of conservation of energy. It states that the total energy of a closed system remains constant. This means that energy can neither be created nor destroyed, only transferred or transformed from one form to another. Hence, in our example of heating the pot in figure 11.1, what happens if we placed a movable wall (led) to cover the pot? From real life experience we notice that the cover moves. If the lid is impermeable, then there will be no matter exchange, and the system remains a closed system. When an object moves, we know that there is a work that has been performed on it. Work is the transfer of energy that results from a force acting over a distance.

> When a force is applied to an object and the object moves, work is done

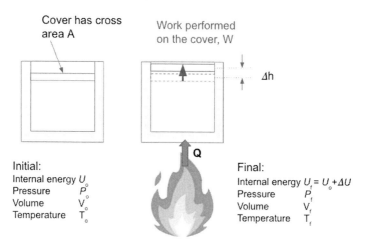

Figure 11.4. Heat from the fire increases the internal energy of the system by the added thermal energy minus the work performed on moving the led.

Work has a sign depending on how it was transferred. It can be

- Positive: Energy is transferred to the system.
- Negative: Energy is transferred out of the system.

So, one could look at this system as in figure 11.4. Fire transfers an amount of thermal energy Q to the system, composed of a pot that contains water and cover. The system initially has an internal energy U_o. When heated, the internal energy of the system is expected to increase by the amount Q. However, part of this energy is transferred out of the system and used as a work, W, to move the cover. Hence, the total amount of increase in the system internal energy is

$$U = Q + W \tag{11.1}$$

Notice that in equation 11.1, we added the work to the thermal energy. This should not be contradicting to the fact that $U < Q$. This is because the work in this case is negative as it is taken from the system. The work is the amount of force applied on the cover to move a distance h, or

$$W = -F \cdot \Delta h \tag{11.2}$$

Again, the sign here is negative as this work is taken from the applied thermal energy. We can rewrite the right-hand side of equation 11.2 as

$$W = -F \cdot \frac{A}{A} \cdot \Delta h = \frac{F}{A} \cdot A\Delta h \tag{11.3}$$

, where A is the area of the cover. Notice that F/A is the force applied on the cover per unit area. That is the pressure, P, as we discussed in the previous unit. The quantity $A\Delta h$ is the change in the system's volume $\Delta \mathbf{V}$. Hence, we can write work as

$$W = -P\Delta \mathbf{V} \tag{11.4a}$$

The change of the volume can be written as the difference between the volume in the final stage compared to the initial stage, $\Delta \mathbf{V} = \mathbf{V}_f - \mathbf{V}_o$.

Example 11.1a:

A thermal energy of 1000 J was applied to a fluid in a closed system with an impermeable movable wall. That caused the volume to increase from 0.05 m³ to 0.055 m³. A pressure of 100 kPa was measured on the wall. What is the amount of change of the internal energy of the system?

Solution:

$$W = -P\Delta \mathbf{V} = (100000 \, Pa) \times (0.055 - 0.05 \, m^3) = -500 \, J$$

From the first law of thermodynamics

$$\Delta U = Q + W = 1000\,J - 500\,J = 500\,J$$

Equation 11.4a assumes that the pressure remains constant during the process. To generalize this concept, one needs to assume that the pressure varies with the volume. Hence, equation 11.4a is valid only over small amount of volume change, $d\mathbf{V}$, over which the pressure could be assumed to remain constant. The amount of work produced over this small change is $dW = -Pd\mathbf{V}$. To calculate the total work, all we need to do is to integrate over the volume

$$W = -\int Pd\mathbf{V} \tag{11.4b}$$

Example 11.1b:

For the case in example 11.1a, if the pressure varies with the volume in the form of $P = 15\ \mathbf{V}^{-2}$ calculate the change in the internal energy of the system.

Solution:

$$W = -\int_{0.05}^{0055} 150\ \mathbf{V}^{-2}dV = [150/\mathbf{V}]_{0.05}^{0.055} = \frac{150}{0.055} - \frac{150}{0.05}$$
$$= 2727 - 3000 = -273\,J$$

$$\Delta U = 1000 - 273 = 727\,J$$

Heat capacity

When heating a substance such as the liquid in figure 11.4, heat capacity defines the amount of thermal energy required to increase the substance temperature by one degree. Mathematically we can write this expression as

$$C = dQ/dT \tag{11.5}$$

It has units of J/K. So, to increase the temperature by T, an amount of thermal energy Q=CT is needed. A normalized form of this constant is known as the specific heat capacity. Specific heat capacity is the heat capacity of the substance per unit mass.

$$c = C/m = \frac{1}{m}dQ/dT \qquad\qquad (11.6)$$

Specific heat capacity has units of $J/K.Kg$ and for different substances it depends on the pressure and temperature. Hence, when looking up values for specific heat capacity for a any substance, one would always find tables that specify the temperature and pressure of the substance. In the scope of this unit, we will not investigate the heat capacity of the different substances. We will rather be content with the mere introduction of this concept as we will focus mainly on ideal gas analysis of thermodynamics.

Example 11.2:

What is the amount of thermal energy needed to boil 1 Kg of water?

Solution:

If we assume that we start from room temperature, then

$$\Delta T = (100^o + 273.15) - (25^o + 273.15) = 75\ K$$

The specific heat capacity for water (at room temperature) is approximately $4.2 \times 10^{-3}\ J/kg \cdot K$.

$$Q = mc\Delta T = 1\ kg \times (4.2 \times 10^{-3}\ J/kg \cdot K) \times (75\ K) = 0.35\ J$$

Ideal gas and temperature

One could notice that in the two examples 11.1a and 11.1b, we considered a compressible fluid. In other words, the thermodynamic process changed the volume of the system. As mentioned in the previous chapter, the compression rate of liquid was assumed to be negligible. Hence, we can state that the fluid in the example was a gas or vapor. Typically, the gas that fills the system comprises of multiple atoms and molecules. Energy exchange occurs due to random collisions between the gas particles. Following a rigorous approach to estimate the change of the system internal energy becomes a complex process. To simplify our analysis, we will restrict our focus to ideal gas. An ideal gas is a theoretical gas composed of large number of identical particles (such as molecules or atoms) that are in constant random motion and interact only through perfectly elastic collisions. Elastic collision indicates no loss of the gas molecule energy due to collision. Hence, the energy of the particles is determined solely by their temperature. Temperature here is measured in Kelvin. The mean kinetic energy can be written as a linear relation with temperature using a proportionality constant $k = 1.38 \times 10^{-23} J/K$. This is known as the Boltzmann coefficient.

$$< \frac{1}{2}mu^2 > = \frac{3}{2}kT \tag{11.7}$$

One can look at the factor 3 here as it is due to the fact that particles can move in three directions. However, the expression in equation 11.7 is a result from quantum mechanics for the average kinetic energy of one atom. The total kinetic energy of the gas molecules, e.g., internal energy, is the multiplication of the number of atoms with the average kinetic energy of one atom.

$$U = N < \frac{1}{2}mu^2 > = \frac{3}{2}NkT \tag{11.8}$$

Here, N is the number of gas atoms inside the volume **V**, at pressure P and temperature T. Typically the number of gas molecules is very large. For that reason, a standard large number of molecules is set and it is referred to as mole. One mole equals $N_o = 6.031023$ atoms. So, to define the total number of atoms N we can conveniently use the number of moles, n, instead. We can say that the total number of atoms is the number of moles multiplied by the number of atoms in one mole, $N = nN_o$. Using this in equation 11.8 we obtain

$$U = \frac{3}{2}nN_o kT = \frac{3}{2}nRT \tag{11.9}$$

Where $R = N_o k = 6.03 \times 10^{23} \times 1.38 \times 10^{-23} = 8.317 \, J/mole.K$. The change of the internal energy then corresponds to the change in the temperature of the gas.

$$\Delta U = \frac{3}{2}nR\Delta T \tag{11.10}$$

Example 11.3.

For the case in example 11.1a, if we assume that 3 moles of gas atoms are in the volume V, then what is the amount of change in the gas temperature

Solution:

$$\Delta U = Q - W = 1000\,J - 500\,J = 500\,J$$

$$\Delta U = \frac{3}{2}nR\Delta T \rightarrow \Delta T = \frac{2\,\Delta U}{3\,nR} = \frac{2}{3}\frac{500\,J}{3\,mole \times 8.317\,J/moleK} \approx 13.3\,K$$

The kinetic theory of gas provides us with another relation between the internal energy of the ideal gas and its pressure and volume as

$$PV = \frac{2}{3}U \tag{11.11}$$

Substituting in equation 11.9, we obtain the relation between the three state variables for ideal gas.

$$PV = nRT \tag{11.12}$$

The derivation of relation in 11.11 is explained in the next section.

Kinetic theory of gas

Let us imagine the following situation where ideal gas molecules are kept in a specific volume **V** where a piston of area A is placed at the right end as shown in figure 11.5. The system is closed, where the piston is impermeable and movable. Gas molecules are always in continuous movement. If we focus our attention on one molecule that has a mass m near the piston then we could assume that at an instant of time, t, it is moving with a velocity $u = (u_x, u_y, u_z)$.

The motion of the molecule would eventually cause a collision with the surface of the piston. If we assume an elastic collision where there is no loss of energy, the molecule will reflect with the exact same velocity. In this case we can examine the momentum of the system before and after the collision as illustrated at the right side in figure 11.5. Here, the piston motion is restricted to the x axis only and hence, we would be concerned only with the x component of the momentum.

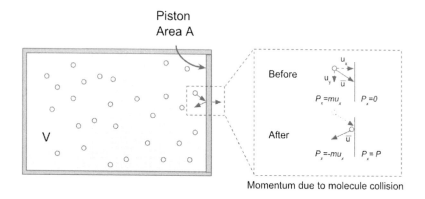

Figure 11.5. Gas molecules inside a closed volume with a piston of area A at one side.

Before collision the piston is assumed to be stationary and hence its momentum is zero. The molecule is moving with a speed u_x along the x axis. Hence the total momentum before collision is

$$p_{before} = mu_x \qquad (11.13a)$$

After collision, the piston will gain a momentum p while the molecule will bounce back with the exam same speed but in the opposite direction, hence

$$p_{after} = -mu_x + p \qquad (11.13b)$$

Applying the conservation of momentum, $p_{before} = p_{after}$ we can estimate the total momentum that is transferred to the piston due to one collision as

$$p = 2mu_x \qquad (11.14)$$

Equation 11.14 describes the momentum gained by the piston due to a collision by one gas molecule. As we explained earlier it is a rather complicated process if we are to consider the effect of each molecule separately to form the total effect. Instead, it would be more reasonable to look at the average effect due to a collection of molecules. For example, we could examine the total momentum transferred to the piston during a time period of dt. If we know that the volume **V** contains N molecules, then it would have a density of $\rho = N/\mathbf{V}$ molecules per unit volume. If all those molecules move with the same velocity, $\bar{u}$, then we could say the total momentum that is transferred to the piston is the multiplication of the momentum due to one collision by the total number of collisions that would occur during the time dt.

The question is how do we know how many collisions would occur during dt? To answer his we would simply assume that all the molecules that are within a distance that is less or equal to $u_x dt$ will reach the piston during the time dt and hence they will collide with the piston. If the piston has an area of A, then we could say that the total number of collisions is the number of molecules within a volume of $\rho A u_x dt$. The total momentum dp achieved during a time interval dt is then

$$dp = (2mu_x) \times (\rho A u_x dt) = 2\rho A m u_x^2 dt \qquad (11.15)$$

Revisiting the definition of momentum and force we recall that

$$\bar{F} = m\bar{a} = m\frac{d\bar{u}}{dt} = \frac{d}{dt}(m\bar{u}) = \frac{d\bar{p}}{dt} \qquad (11.16)$$

When there is no change in the mass, the force is the time derivative of the momentum, and hence we can say that the total force exerted on the piston along the x direction is

$$\bar{F} = d\bar{p}/dt = 2\rho A m u_x^2 \qquad (11.17)$$

The relation in 11.17 is derived for piston of a specific area A. It would be more practical here to consider the force per unit area, or pressure, that is applied on the piston.

$$P = F/A = 2\rho m u_x^2 \qquad (11.18)$$

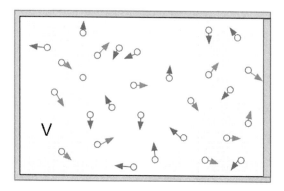

Figure 11.6. Random motion of the gas molecules inside a volume V. The graph
shows that not all the molecules will contribute to the collision with the piston at the
right side.

One issue that remains in equation 11.18. It is the fact that we assumed all
molecules to travel at the same velocity. In reality, gas molecules are
moving in all directions with almost equal probability as demonstrated in
figure 11.6.

If we assume uniform distribution of the direction of motion, we could
safely say that half of the molecules will have a velocity with the x-
component heading towards the piston, however the other half does not as
it is graphically illustrated in figure 11.6 where the red arrows (color online)
are the molecules that would contribute in hitting the movable wall or the
piston. In addition, every molecule would as well have its own velocity and
what we would be interested in is the average of the square of these
velocities or $< u_x^2 >$. The average pressure on the piston is then

$$P = \rho m < u_x^2 >$$ (11.19)

If we assume that the molecules could move equally in every possible
direction, then the averages of their velocity components in x, y and z should
be equal. In other words, $< u_x^2 > = < u_y^2 > = < u_z^2 >$. The average of the
square amplitude of the velocity is

$< u^2 > = < u_x^2 + u_y^2 + u_z^2 > = < u_x^2 > + < u_y^2 > + < u_z^2 > = 3 <
u_x^2 >$. In other words, we could say that $< u_x^2 > = \frac{1}{3} < u^2 >$. Equation
11.18 can then be written as

$$P = \frac{1}{3}\rho m < u^2 > \tag{11.20}$$

If we multiply both sides by the volume **V** and move the mass inside the average operation, equation 11.20 becomes

$$P\mathbf{V} = \frac{2}{3}N < \frac{1}{2}mu^2 > \tag{11.21}$$

The term $< \frac{1}{2}mu^2 >$ is the average molecule's kinetic energy. The total kinetic energy inside the volume can be approximated by the multiplication of the number of molecules N by the average kinetic energy of each molecule, $U = N < \frac{1}{2}mu^2 >$. Hence, equation 11.21 becomes

$$P\mathbf{V} = \frac{2}{3}U \tag{11.22}$$

We need to keep in mind that in the derivation of equations 11.21 and 11.22 we assumed that the volume contains only one type of gas. That is why the mass was assumed to be constant for all molecules. Also, we have neglected collisions between molecules inside the volume. Hence the energy U is assumed to be the total internal energy of the gas inside the volume.

Changes in the state variables

As we mentioned in the introduction, all or some of the state variables change when a thermodynamic process is applied. In table 11.4 we summarized four thermodynamic processes depending on their effect on the variables and energy transfer. In this section, we will try to state the effect of these processes using the first law of thermodynamics and the ideal gas equation, $P\mathbf{V} = nRT$ as in 11.12. To estimate the variation in the state parameter we can apply the small perturbation theory where all the three parameters are changed by small amounts: ΔP, $\Delta \mathbf{V}$ and ΔT. Hence, we can write the perturbed form of equation 11.12 as

$$(P + \Delta P)(\mathbf{V} + \Delta \mathbf{V}) = nRT(T + \Delta T) \tag{11.23a}$$

Expanding the terms in both side

$$P\mathbf{V} + P\Delta \mathbf{V} + \Delta P\mathbf{V} + \Delta P\Delta \mathbf{V} = nRT + nR\Delta T \tag{11.23b}$$

We know that $P\mathbf{V} = nRT$, then we can simplify 11.32b to

$$P\Delta \mathbf{V} + \Delta P\mathbf{V} + \Delta P\Delta \mathbf{V} = nR\Delta T \tag{11.24}$$

Here, we assume that the changes are very small. Hence, the product $\Delta P \Delta V$ can be neglected and the change of the state parameters can be written as.

$$P\Delta V + \Delta PV = nR\Delta T \tag{11.25}$$

The change is the difference between the final state and initial state of equilibrium. By that it means a state when the system does not undergo any further changes in time. For instance, $\Delta P = P_f - P_o$, where P_o and P_f are system pressure before and after the thermodynamic process. If the changes of the state parameters are infinitesimal, equation 11.25 can be re-written in its differential form as

$$dPV + PdV = nRdT \tag{11.26}$$

Now, we are ready to examine the different processes in table 11.4 and define them mathematically.

Table 5.5. The governing equations for different thermodynamic processes.

Process	Implication	Governing equations
Isothermal	T is constant $dT = 0$	$dPV + PdV = 0 \rightarrow \frac{dP}{P} + \frac{dV}{V} = 0$ Integration: $\ln(P) + \ln(V) = const.$ $\ln(PV) = const. \rightarrow PV = const.$ $\Delta U = \frac{3}{2}nRT = 0 = W + Q \rightarrow W = -Q$
Adiabatic	$Q = 0$	$dPV + PdV = nRdT$ $\Delta U = W = \frac{3}{2}nRT$
Isobaric	P is constant $dP = 0$	$PdV = nRdT$ Integration: $V = \frac{nR}{P}T$ $\Delta U = Q + W = \frac{3}{2}nRT$
Isochoric	V is constant $dV = 0$ $W = 0$	$dPV = nRdT$ Integration: $P = \frac{nR}{V}T$ $U = Q = \frac{3}{2}nRT$

Example 11.4:

Calculate the temperature of 4 moles of ideal gas that is placed in a closed system with fixed, impermeable and diathermal walls with volume of 0.05 m^3 if the pressure inside is measured to be 300 kPa.

Solution:

$$PV = nRT \rightarrow T = PV/nR = \frac{300000 \; Pa \times 0.05 \; m^3}{3 \; moles \times 8.317 \frac{J}{K.mole}}$$

$T = 601.2 \; K$. Which is $601.2 - 273.15 = 328^o \; C$

Example 11.5:

If the gas temperature in example 5.4 increased by 30 degrees, what is the change in the gas pressure and volume? What is the change in the internal energy of the gas?

Solution:

The process of increasing the temperature exists in fixed, diathermal and impermeable walls. That means the volume does not change, which indicates an isochoric process.

$$\Delta V = 0 \rightarrow W = 0$$

$$P = \frac{nR}{V} T \text{ and } P + \Delta P = \frac{nR}{V}(T + \Delta T) \rightarrow \Delta P = \frac{nR}{V} T$$

$$\Delta P = \frac{nRT}{V} = \frac{3 \; moles \times 8.317 \frac{J}{K.mole} \times 30 \; K}{0.05 \; m^3} = 14970.6 \text{ Pa}$$

The change in the internal energy is

$$\Delta U = \frac{3}{2} nR\Delta T = \frac{3}{2} \times 3 \; mole \times 8.317 \frac{J}{K.mole} \times 30 \; K \approx 1123 \; J$$

Example 11.6:

What is the work performed on a movable wall of a closed system that contains 3 moles of ideal gas that changes its temperature from $50°$ C to $300°$ C. The walls are impermeable and adiabatic.

Solution:

The walls are adiabatic hence there is no energy transfer in or outside the system. Hence, this is an adiabatic process, and the change of the internal energy equals the work.

$$\Delta U = \frac{3}{2}nRT = 323 \ mole \times 8.317 \frac{J}{K.mole} \times (300 - 50) \approx 28069 \ J$$

$$W = \Delta U = 28069 \ J$$

More on the adiabatic process

In an adiabatic process, work depends only on the temperature change, $W = \frac{3}{2}nR\Delta T$. Let us consider the work performed due to an infinitesimal increase in temperature dT, $dW = \frac{3}{2}nRdT$. The work can be as well written as

$$dW = -PdV = \frac{3}{2}nRdT \tag{11.27}$$

From table 11.5 we know that $dT = (PdV + VdP)/nR$. Then,

$$-PdV = \frac{\frac{3}{2}nR(PdV+VdP)}{nR} \tag{11.28a}$$

Arranging the terms on both sides,

$$\frac{5}{2}PdV + \frac{3}{2}VdP = 0 \rightarrow \frac{dP}{P} + \frac{5}{3}\frac{dV}{V} = 0 \tag{11.28b}$$

Integrating equation 11.28b we obtain the following relation

$$\ln(P) + \ln(V^{5/3}) = constant \tag{11.29a}$$

or

$$PV^{5/3} = constant \tag{11.29b}$$

We can as well rearrange 11.29b as

$$PV\left(V^{3/2}\right) = constant \tag{11.29c}$$

We know that for the ideal gas, $PV = nRT$, hence we can write the following

$$T\,V^{2/3} = constant \tag{11.29d}$$

In equation 11.29d we neglected the term nR as it is constant. So, for an adiabatic process between two states A and B, we can write the $P_A V_A^{5/3} = P_B V_B^{5/3}$ or $T_A V_A^{2/3} = T_B V_B^{2/3}$.

11C. The second law of thermodynamics

The second law of thermodynamics states that the total entropy of a closed system (or an isolated system) always increases over time or remains constant in idealized reversible processes. That means that the level of disorder or randomness in the system also increases.

Entropy

Entropy can be defined as a measure of the disorder or randomness of a system. It represents the amount of energy in a system that is not available to do useful work. In simpler terms, entropy can be thought of as a measure of the level of chaos or randomness and hence the more order the system has the lower the entropy.

To get a better understanding, let us consider the following example. If we arrange the library bookshelves in the morning neatly where books are arranged in alphabetical order and the shelves are sorted by topic then we could say that the library has a high order and thus low entropy. Now, let the students start using the library where they move the books around and misplace them on the shelves. We could now conclude that the order of the library has decreased, and the randomness has increased. In thermodynamics terms, we would say that the entropy of the library as a system has increased. Also, as long as the library doors remain open and students keep on using it, the entropy would keep on increasing. Now, let us close the doors (evening time) and make sure that there is no external wind or disturbance that would occur in the library. Also, we hired a person to ensure that he compensates for any changes that could happen during that period to reverse its effect. In this case, the entropy of the system would remain unchanged. To bring this entropy to its original low value, one would need to do work to re- arrange the books back again.

Similarly, in thermodynamics, we could imagine the library as a closed system where in any process where energy is transformed or transferred from one form to another, some of it is inevitably lost as unusable heat energy, which increases the overall disorder of the system. The entropy of the system would then always keep on increasing. Mathematically, we can write the change of the entropy as follows

$$\Delta S = Q/T \qquad (11.30)$$

Where Q is the amount of thermal energy given or released from the system and T is the temperature (in Kelvin) at which the process occurs. The value of S should be greater or equal zero. When a process on the system results in an increase in the system entropy (randomness) then the process is called irreversible. In this process there is a finite amount of energy Q that is spread or dissipated in the system which increases the randomness of the system. If on the other hand, a process is performed in a way that kept the system always at thermal equilibrium, then $Q = 0$ and the entropy remains constant. This process is referred to as reversible. An example of a reversible process is a gas expanding isothermally against a piston, where the pressure is gradually decreased so that the gas expands slowly enough to maintain thermal equilibrium with its surroundings. This process can be reversed by compressing the gas slowly back to its original state, and the system and surroundings will return to their original state.

Table 11.6. Equilibrium

Thermal equilibrium	When two or more objects are in contact with each other and have no net transfer of heat between them. This happens when the objects are at the same temperature.
Mechanical equilibrium	When the sum of all the forces acting on an object or system is zero. In other words, the net pressure inside the system equals the surrounding.

Heat transfer

Let us consider the example of a hot cup of tea that is placed inside a closed cool room. As we predict from our real-life experience, the tea will start to cool down transferring thermal energy to the surrounding (increasing the surrounding temperature) till an equilibrium is reached where both the cup and surrounding are at the same temperature. Increasing the temperature of the surrounding increases the kinetic energy of surrounding gas molecules and hence decreases the predictability of localizing the molecules in the system (the cup inside the room). This increase in the randomness increases the total system entropy and the process is irreversible.

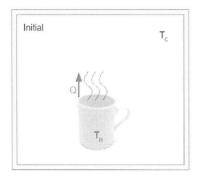
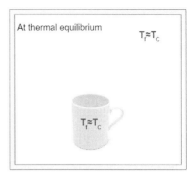

Figure 11.7. Cooling a hot cup of tea in a closed room increases the entropy of a system composed of the room and the cup inside

The final temperature as illustrated in figure 11.7 is almost the same as that of the room at the initial stage as the room volume is much larger than that of the cup. However, we are certain that a certain amount of thermal energy has transferred from the cup to the surrounding volume. If we assume hypothetically that the walls of the teacup are adiabatic (no thermal transfer), then the only surface that causes heat transfer is the cup surface. Heat transfer occurs then through the means of convection from the top air molecule (in contact with the hot liquid) to the surrounding molecules. Also, heat is transferred inside the liquid to the top layer through convection as well. The amount of heat transfer through convection can be expressed as

$$Q = h\,A\Delta T \tag{11.31}$$

Here, h is the overall thermal transfer coefficient, A is the area of the surface and T is the difference between the initial and final temperatures of the liquid in the cup. The thermal coefficient in our case is a combination of both outside convection by air molecules, h_o, and inside convection by the liquid, h_i. This can be written as

$$\frac{1}{h} = \frac{1}{h_i} + \frac{1}{h_o} \tag{11.32}$$

Example 11.6:

If the teacup was originally at temperature $60°$ and the room is at room temperature, then estimate the change in the entropy? The convection thermal transfer coefficients of air and water are 50 J/m^{2}K and 200 J/m^{2}K respectively and the cup has a radius of 4 cm.

Solution:

$T_H = 60^o$ and $T_C = 25^o$ (Room temperature), The final temperature $T_f \approx 25^o$.

$\Delta T = 60 - 25 = 35^o$.

$\frac{1}{h} = \frac{1}{50} + \frac{1}{200} \rightarrow h = 40 \, J/m^2K$

$A = \pi(0.04)^2 = 0.005 \, m^2$

$Q = hA\Delta T = 40 \times 0.005 \times 35 = 7 \, J$

The change in the entropy $\Delta S = Q/T_f = 7/(25 + 273.15) = 0.0235 \, J/K$

Now if we placed a cup of iced tea in a closed room at room temperature, that what would be expected to happen? We know the answer to this, the iced tea will start to increase its temperature while the surrounding reduces its temperature till a thermal equilibrium is reached. Again, this process involves increasing the randomness of the liquid molecules. You can think of that as when we melt a cube of ice inside the iced tea, transformation from an organized solid to a less organized liquid indeed increases the system randomness. Hence, again the entropy of the system increases, and the process is irreversible. In other words, you cannot form the ice cube back again using a process that keeps thermal equilibrium with the surroundings.

Example 11.7:

If the teacup in example 5.6 contains an iced tea instead with an initial temperature 4° and the room is at room temperature, then estimate the change in the entropy?

Solution:

$T_H = 25^o$ and $T_C = 4^o$, The final temperature $T_f \approx 25^o$

$\Delta T = 25 - 4 = 21^o$.

$h = 40\,J/m^2K$

$A = 0.005\,m^2$

$Q = hA\Delta T = 40 \times 0.005 \times 21 = 4.2\,J$

The change in the entropy $\Delta S = Q/T_f = 4.2/(25 + 273.15) = 0.0141\,J/K$

As we could see in both examples heat has always transferred from the hotter to the cooler. In other words, we can re-state the second law of thermodynamics as:

> Heat transfer occurs spontaneously from higher to lower temperature bodies but never spontaneously in the reverse direction.

We can then restate the law as: it is impossible that any process can solely cause heat transfer from cooler to hotter.

Heat engine

As heat always transfers from hotter to cooler, one can possibly transfer part of this energy into work (mechanical work). This concept is illustrated in figure 11.8.

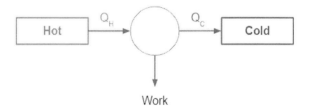

Figure 11.8. Heat transfer is used to produce work.

If we return to the example of the hot teacup, we will realize that heat transfer stops once equilibrium is reached. Hence, to be able to extract work from the heat transfer in figure 11.8, both hot and cold regions need to remain at their temperatures. In other words, we need to make sure that we have an external means to keep the hot body at its temperature (a heater, oven or a boiler for example), which is commonly called a furnace. The cool body is kept cold at a constant temperature (through convection by cooling liquid or radiation to the air), which is commonly referred to as refrigerator.

If we assume that we manage to keep both hot and cold bodies using external means, then part of the produced heat is transferred into work. However, to achieve mechanical work, the volume of the system would need to change during the process to ensure motion. An expandable fluid, such as vapor or gas, is required to fill this system. To simplify the analysis let us assume an ideal gas that fills a diathermal and impermeable cylinder with a piston that can move up and down (movable closed system).

Using the first law of thermodynamics and if the change in the internal energy occurs only in the thermal energy of the system, $\Delta U = U_f - U_o$, where U_o is the initial system thermal energy before heating, we can write the following relation

$$Q_f - Q_o = Q_{in} + W - Q_{out} \qquad (11.33)$$

Here, Q_{in} is the input thermal energy and Q_{out} is the thermal energy released from the system. We can write the hot body thermal energy as the summation of the initial energy to the one that is added, assuming no loss, $Q_H = Q_o + Q_{in}$. The cold body thermal energy can be written as the sum of the final energy and that which is released, $Q_C = Q_f + Q_{out}$. Hence, we can rearrange equation 11.33 as

$$W = Q_H - Q_C \qquad (11.34)$$

The efficiency of the system is the obtained work divided by the hot body thermal energy.

$$Eff = \frac{W}{Q_H} = 1 - \frac{Q_C}{Q_H} \qquad (11.35)$$

Example 11.7:

If the thermal energy provided by the hot body (furnace) is 17000 J and that released by the cold body is 11000 J. What is the efficiency of the engine?

Solution:

The efficiency of the engine is

$$Eff = 1 - \frac{Q_C}{Q_H} = 1 - \frac{11000}{17000} = 0.353$$

Carnot Cycle

The idealistic heat engine described in the previous section is known as Carnot heat engine. It is a theoretical engine that operates on a reversible thermodynamic cycle and is often used as a standard for comparison with actual engines. It was first proposed by French engineer Sadi Carnot in 1824 and is sometimes referred to as the Carnot cycle.

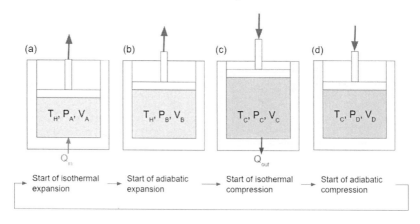

Figure 11.9. An idealistic heat engineer following the Carnot cycle.

The engine comprises four processes as shown in figure 11.9 and listed below:

1. In the initial stage, the furnace provides thermal energy Q_{in} that keep the system at temperature T_H. During this stage, the piston

starts to move, and the volume expands while keeping the temperature constant, an **isothermal expansion** process. During this process

    a. $P_A V_A = constant = nRT_H = P_B V_B$

    b. $W_{AB} = -\int_A^B P dV = nRT_H \ln\left(\frac{V_A}{V_B}\right) = -Q_{in}$

2. In this process the system continues to expand, however with no heat supply and in this case the temperature starts to cool down following an **adiabatic expansion** process. The process is adiabatic as there is no energy transfer to or outside the system. During this process

    a. $P_B V_B^{5/3} = constant = P_C V_C^{5/3}$ or $P_B T_B^{2/3} = constant = P_C T_C^{2/3}$

    b. $W_{BC} = \frac{3}{2} nR(T_C - T_H)$

3. Here, the cooled gas starts to compress. This compression would typically increase the temperature. However, the system is kept at a constant temperature by leaking its excess thermal energy, Q_{out}, to a refrigerator. Hence it follows an **isothermal compression** process.

    a. $P_C V_C = constant = nRT_C = P_D V_D$

    b. $W_{CD} = -\int_C^D P dV = nRT_C \ln\left(\frac{V_C}{V_D}\right) = -Q_{out}$

4. In the final process, the volume continues to be compressed but with no outlet to cool the system following an **adiabatic compression** process. The pressure increases while the volume is decreasing. The temperature rises till we reach the original state variables: P_A, V_A and T_H.

    a. $P_D V_D^{5/3} = constant = P_A V_A^{5/3}$ or $P_D T_D^{2/3} = constant = P_A T_A^{2/3}$

    b. $W_{DA} = \frac{3}{2} nR(T_H - T_C)$

Using the ideal gas equation in 11.11 we can write the pressure as a function of volume as $P(V, T) = \frac{nRT}{V}$. We can then construct plots for the pressure as a function volume (**PV** diagrams) at the two temperatures, T_H and T_C as illustrated in figure 11.10.

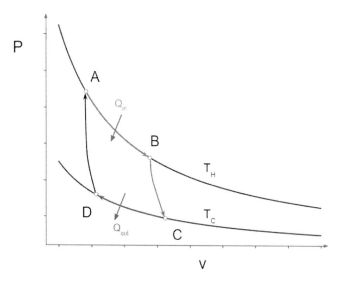

Figure 11.10. The PV diagram for the Carnot cycle.

The plots show the reversible Carnot cycle starting from an initial state A.

1. *From A→B*: The system follows an isothermal expansion (following the red curve) where volume is increased, and temperature is kept constant. Remember that the two lines marked T_H and T_C represent the pressure as a function of volume at constant temperatures.
2. *From B→C*: The system follows an adiabatic expansion where all the three state variables change with no energy exchange between the system and the outside.
3. *From C→D*: The volume is reduced while the temperature is kept constant by leaking its excess thermal energy to the environment. That follows an isothermal compression process.
4. *From D→A*: The system returns to the original state following an adiabatic compression process.

Total work produced by the engine

The total work produced by this process is the sum of all the works

$$W = W_{AB} + W_{BC} + W_{CD} + W_{DA} \tag{11.34}$$

From the Carnot cycle's list, we introduced earlier, we can write the work as

$$W = nRT_H \ln\left(\frac{V_A}{V_B}\right) + \frac{3}{2}nR(T_C - T_H) + nRT_C \ln\left(\frac{V_C}{V_D}\right) + \frac{3}{2}nR(T_H - T_C) \tag{11.35}$$

Notice that $W_{BC} = -W_{DA}$ and hence equation 11.35 can be simplified as

$$W = nR\left[T_H \ln\left(\frac{V_A}{V_B}\right) + T_C \ln\left(\frac{V_C}{V_D}\right)\right] \tag{11.36}$$

We know that $V_B > V_A$, then we can write equation 11.36 as

$$W = -nR\left[T_H \ln\left(\frac{V_B}{V_A}\right) - T_C \ln\left(\frac{V_C}{V_D}\right)\right] \tag{11.37}$$

Notice that we pulled the minus sign outside as the total work is expected to be taken from the engine. Recalling the relation between temperature and volume in equation 11.29d, we can write the following

$$V_B/V_C = (T_C/T_H)^{3/2} = V_A/V_D \tag{11.38a}$$

Which indicates that

$$\frac{V_B}{V_A} = \frac{V_C}{V_D} \tag{11.38b}$$

Using the expression in 11.38b into 11.37 we obtain

$$W = -nR \ln\left(\frac{V_B}{V_A}\right)[T_H - T_C] \tag{11.39}$$

If we define $Q_H = nR \ln\left(\frac{V_B}{V_A}\right)T_H$ and $Q_C = nR \ln\left(\frac{V_B}{V_A}\right)T_C$ then we obtain the relation in 11.34 except for the negative sign, which as explained earlier an indication that the work was taken from the system. In this case the efficiency of the engineer can be written as

$$Eff = 1 - \frac{Q_C}{Q_H} = 1 - \frac{T_C}{T_H} \tag{11.40}$$

Example 11.6:

A Carnot engine is designed to keep the hot body at 200°C and the cold body at 30°C. Calculate the different cycle state variables and total work obtained if the initial volume of the cylinder containing 3 moles of ideal gas is 0.01 m³. Notice that during the adiabatic process, the volume expanded to 0.02 m³.

Solution:

Given:

$T_H = 200° + 273.15 = 473.15\ K$ and $T_C = 30° + 273.15 = 303.15K$

$V_A = 0.01\ m^3$ and $V_B = 0.02\ m^3$

$n = 3\ moles$

1. *For the isothermal expansion process:*

$P_A V_A = nRT_H \rightarrow P_A = 3\ moles \times 8.317 \frac{J}{mole} \times 473.15 \frac{K}{0.01} m^3$

$\rightarrow P_A = 1180556\ Pa$

$W_{AB} = -nRT_H \ln\left(\frac{V_B}{V_A}\right) = -3 \times 8.317 \times 473.15 \times \ln\frac{0.02}{0.01} = -8183\ J$

$Q_{in} = -W_{AB} = 8183\ J$

$P_A V_A = P_B V_B \rightarrow P_B = \frac{P_A V_A}{V_B} = 590278\ Pa$

2. *For the adiabatic expansion process:*

$P_C V_C = nRT_C$

$T_C V_C^{2/3} = T_H V_B^{2/3}$

$$\mathbf{V_C} = \mathbf{V_B} \left(\frac{T_H}{T_C}\right)^{3/2} = 0.02 \times \left(\frac{473.15}{303.15}\right)^{3/2} = 0.039\ m^3$$

$$P_C = nRT_C/\mathbf{V_C} = 3 \times 8.317 \times 303.15/0.039 = 193956\ Pa$$

$$W_{BC} = \frac{3}{2}nR(T_H - T_C) = \frac{3}{2}3 \times 8.317 \times 170 = 6362.5\ J$$

3.   For the isothermal compression process:

$$P_D\mathbf{V_D} = P_C\mathbf{V_C}$$

$$P_D\mathbf{V_D} = nRT_C$$

4.   For the adiabatic compression process:

$$T_C\mathbf{V}_D^{2/3} = T_H\mathbf{V}_A^{2/3}$$

$$\mathbf{V_D} = \mathbf{V_A} \left(\frac{T_H}{T_C}\right)^{3/2} = 0.01 \times \left(\frac{473.15}{303.15}\right)^{3/2} = 0.0195\ m^3$$

$$P_D = nRT_C/\mathbf{V_D} = 3 \times 8.317 \times 303.15/0.0195 = 387912\ Pa$$

$$W_{CD} = nRT_C \ln\left(\frac{\mathbf{V_C}}{\mathbf{V_D}}\right) = 5243\ J$$

$$W_{DA} = \frac{3}{2}nR(T_C - T_H) = -\frac{3}{2}3 \times 8.317 \times 170 = -6362.5\ J$$

The total work

$$W = W_{AB} + W_{BC} + W_{CD} + W_{DA}$$

$$W = -8183\ J + 6362.5J + 5243\ J - 6362.5J = -2940\ J$$

The efficiency of the engine is

$$Eff\ =\ 1 - \frac{T_C}{T_H} = 1 - \frac{303.15}{473.15} = 0.306$$

This process is shown in figure 11.11.

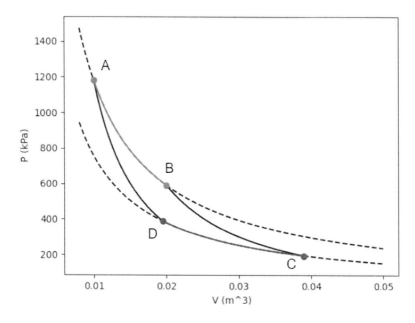

Figure 11.11. The PV diagram of the Carton cycles in example 11.6.

11.D. Third law of thermodynamics

The third law of thermodynamics states that as the temperature of a system approaches absolute zero (0 Kelvin or -273.15 degrees Celsius), the entropy of the system approaches a minimum value or a constant value if the system is perfectly ordered. In other words, if a system could somehow be cooled down to absolute zero, all motion within the system would cease, and the entropy of the system would reach a minimum value. The third law of thermodynamics is particularly important for studying the behavior of materials at very low temperatures, as it helps us understanding how the entropy of a system changes as it approaches absolute zero.

11.E. Summary

- Thermodynamics deals with the relationships between heat, energy and work.
- Internal energy, U
    - It deals with the microscopic state of the particles
        - Potential energy:
            - Location
            - Chemical properties
            - Configuration
        - Kinetic energy:
            - Motion of the particles
            - It is known as thermal energy
            - Temperature:
                - Measure of average kinetic energy
                - Units of Kelvin (K)
                - Kelvin = 273.15 + Celsius
                - Absolute zero = zero kinetic energy
- Heat
    - Transfer of energy between two objects due to a temperature difference
        - Conductive
            - Direct contact between objects
        - Convection
            - Movement of fluid
        - Radiation
            - Emission of electromagnetic waves
- Thermodynamic systems
    - Open system
        - Exchange of matter and exchange of energy
    - Closed system
        - Exchange of energy
    - Isolated system
        - No exchange of energy or matter.

- Mobility
  - Fixed
    - Walls do not move
  - Movable
    - Walls can move
- Exchange of matter
  - Permeable
    - Allows transfer of matter
  - Impermeable
    - Do not allow transfer of matter
- Exchange of energy
  - Diathermal
    - Allows energy transfer
  - Adiabatic
    - Does not allow energy transfer
- State variables
  - Volume: $\mathbf{V}$
  - Pressure: P
  - Temperature: T
- Thermodynamic process
  - Based on energy applied
    - Mechanical process
      - Application of mechanical work
    - Thermal process
      - Application of thermal energy
    - Chemical process
      - Matter exchange
  - Based on the effect on state variables
    - Isothermal process
      - $\Delta T = 0$
    - Adiabatic process
      - $\Delta U = 0$
    - Isobaric process
      - $\Delta P = 0$
    - Iscochoric process
      - $\Delta \mathbf{V} = 0$

- First law of thermodynamics
  - Work (W)
    - $\Delta W = -P\Delta \mathbf{V}$
    - $W = -\int P d\mathbf{V}$
    - Positive work
      - Energy is transferred to the system
    - Negative work
      - Energy is transferred out of the system
    - $U = Q + W$
  - Heat Capacity (C)
    - $C = dQ/dT$
    - Units J/K
    - Per unit mass
      - $c = C/m = \frac{1}{m} dQ/dT$
- Ideal gas
  - Theoretical gas composed of a large number of identical particles
  - In constant random motion
  - Interact only through perfectly elastic collisions
  - The internal energy
    - $U = \frac{3}{2} nRT$
      - n is the number of moles.
      - $R = 8.317\, J/mole.K$
    - $P\mathbf{V} = \frac{3}{2} nRT$
  - Change of state variables
    - $P\mathbf{V} + P d\mathbf{V} = nRdT$
    - Isothermal
      - $P\mathbf{V} = const.$
      - $W = -Q$
    - Adiabatic
      - $P\mathbf{V}^{5/3} = const.$
      - $Q = 0$
    - Isobaric
      - $P = const.$
      - $\mathbf{V} = \frac{nR}{P} T$

- Isochoric
  - $\mathbf{V} = const.$
  - $P = \frac{nR}{\mathbf{V}} T$
  - $W = 0$
- The second law of thermodynamics
  - Entropy
    - A measure of the disorder or randomness of a system
    - $\Delta S = \frac{Q}{T}$
  - Thermal equilibrium
    - No heat transfer between two objects
  - Mechanical equilibrium
    - The net pressure inside the system equals the surrounding.
  - Heat transfer through convection
    - $Q = h \, A \Delta T$
      - h is the overall thermal transfer coefficient
        - $\frac{1}{h} = \frac{1}{h_i} + \frac{1}{h_o}$
        - h_i: Thermal coefficient of liquid
        - h_o: Thermal coefficient of air
      - A is the area of the surface
- Heat engine
  - $W = Q_H - Q_C$
  - Efficiency
    - $Eff = \frac{W}{Q_H} = 1 - \frac{Q_C}{Q_H}$
  - Carnot cycle
    - Isothermal expansion
      - $P_A \mathbf{V}_A = P_B \mathbf{V}_B$
    - Adiabatic expansion
      - $P_B \mathbf{V}_B^{5/3} = P_C \mathbf{V}_C^{5/3}$
    - Isothermal compression
      - $P_C \mathbf{V}_C = P_D \mathbf{V}_D$
    - Adiabatic compression
      - $P_D \mathbf{V}_D^{5/3} = P_A \mathbf{V}_A^{5/3}$

- Total work
  - $$W = -nR \ln\left(\frac{V_B}{V_A}\right)[T_H - T_C]$$
- Efficiency
  - $$Eff = 1 - \frac{Q_C}{Q_H} = 1 - \frac{T_C}{T_H}$$

11.F. Review questions

1. A system that allows only energy exchange is called

    a. Open system
    b. Closed system
    c. Isolated system.

2. A wall that does not move but allows matter and heat exchange is

    a. Fixed, permeable and a adiabatic
    b. Movable, impermeable and diathermal
    c. Fixed, permeable and diathermal

3. Which process of these does not change pressure

    a. Adiabatic
    b. Isobaric
    c. Isochoric

4. Which of the following is not a state variable

    a. Pressure   b. Temperature c. Heat d. Volume

5. When warming around the camping fire, the heat reaches or hand via

    a. Thermal conduction
    b. Radiation
    c. Thermal conduction

6. When boiling water in a pot with cover, not all the heat is transferred into rising the water temperature. Where did the rest of the thermal energy go

    a. Emitting light
    b. Performing work moving the cover
    c. Returning back to the fire

7. The application of 10 J of thermal energy increased the internal
    energy of a gas inside a closed system with movable impermeable
    walls by 8 J. If the pressure remains at 1 atmosphere before and
    after the process, what the resultant change in volume?
    *Hint: W = Q - DU = 10 - 8 = 2J. W = PDV—> DV=W/P=2
    J/100000 Pa = 0.0002 m3 = 20 cm3*

8. 3 moles of ideal gas are placed in a closed system and kept at 60
    degrees Celsius at atmospheric pressure. What is the volume of the
    system?

9. A thermal engine operates between high and low temperatures of
    200 and 60 degrees Celsius. Calculate the engine efficiency.

10. In an isothermal process, the initial volume was 0.02 m3. If the
    final volume and pressure were 0.03 m3 and 120000 Pa, what was
    the intimate pressure in Pa?

11. If the process in the previous question was adiabatic, what is the
    initial pressure in Pa?

12. What is the high temperature in the previous process if 1 mole of
    an ideal gas was used? (In Celsius)

13. In a Carnot cycle, if the ratio of the volumes VA/VB = VC/VD =
    2, what is the produced work if the high and low temperatures are
    50 and 150 degrees Celsius and the system contains 2 moles of an
    ideal gas.

REFERENCES

[1] Richard P. Feynman, Robert B. Leighton, Matthew Sands, The Feynman Lectures on Physics, (https://www.feynmanlectures.caltech.edu)

[2] Ali Sadegh, William Worek, Marks' Standard Handbook for Mechanical Engineers, 12th Edition, McGraw Hill; 12th edition (November 8, 2017)

[3] Edward J. Shaughnessy, Jr., Ira M. Katz, James P. Schaffer, "Introduction to fluid mechanics," New York Oxford, Oxford University Press (2005)

[4] Crandall, Dahl, Lardner (1959). An Introduction to the Mechanics of Solids. Boston: McGraw-Hill. ISBN 0-07-013441-3.

[5] CRC Handbook of Chemistry and Physics, 99th Edition (Internet Version 2018), John R. Rumble, ed., CRC Press/Taylor & Francis, Boca Raton, FL.

[6] Fellows, P.J. (2009), Food Processing Technology: Principles and Practice (3rd ed.), Woodhead Publishing, ISBN 978-1845692162

[7] Bruce R. Munson, Donald F. Young, Theodore H. Okiishi, Wade W. Huebschf, "Fundamentals of Fluid Mechanics sixth Edit," John Wiley & Sons Inc., (2009)

[8] Pallas, N.R. and Harrison, Y, Colloids and Surfaces (1990) 43,169–194

[9] Jean-Philippe Ansermet, Sylvain D. Brechet, Principles of Thermodynamics 1st Edition, Cambridge University Press; 1st edition (February 28, 2019)

[10] H. W. Liepmann, A. Roshko, "Elements of gas dynamics," Dovers publications (2002)